Industrial Minerals Flotation—Fundamentals and Applications

Industrial Minerals Flotation—Fundamentals and Applications

Guest Editors

Xuming Wang
Jan D. Miller

MDPI

Basel • Beijing • Wuhan • Barcelona • Belgrade • Novi Sad • Cluj • Manchester

Guest Editors

Xuming Wang
Department of Materials
Science and Engineering
University of Utah
Salt Lake City
USA

Jan D. Miller
Department of Materials
Science and Engineering
University of Utah
Salt Lake City
USA

Editorial Office
MDPI AG
Grosspeteranlage 5
4052 Basel, Switzerland

This is a reprint of the Special Issue, published open access by the journal *Minerals* (ISSN 2075-163X), freely accessible at: https://www.mdpi.com/journal/minerals/special_issues/25N2Y6BO5K.

For citation purposes, cite each article independently as indicated on the article page online and as indicated below:

Lastname, A.A.; Lastname, B.B. Article Title. *Journal Name* **Year**, *Volume Number*, Page Range.

ISBN 978-3-7258-5617-6 (Hbk)
ISBN 978-3-7258-5618-3 (PDF)
https://doi.org/10.3390/books978-3-7258-5618-3

Contents

About the Editors

Xuming Wang

Xuming Wang is a Research Professor in the Department of Materials Science and Engineering at the University of Utah. His research focuses on mineral processing, coal preparation, hydrometallurgy, and energy materials. He has made significant contributions to flotation chemistry, critical mineral recovery and recycling, and surface chemistry, with expertise in advanced characterization techniques such as sum-frequency generation (SFG) spectroscopy, FTIR, and atomic force microscopy (AFM). In recent years, he has expanded his work to address critical mineral recovery from diverse resources and to develop next-generation solid polymer electrolytes for lithium-ion and all-solid-state lithium–sulfur batteries. His interdisciplinary research bridges mineral processing with sustainable energy technologies. He serves as an Editor of the *Journal of Physicochemical Problems of Mineral Processing* and is on the Editorial Board of *Carbon Neutral Systems*. He was recognized as the SME Henry Krumb Lecturer (2013–2014) and received the Surface Innovation Prize (2016) from the Institution of Civil Engineers.

Jan D. Miller

Jan D. Miller is the Ivor D. Thomas Distinguished Professor of Metallurgical Engineering at the University of Utah. He has served on the University of Utah faculty for 50 years and is a member of both the National Academy of Engineering (NAE) and the National Academy of Inventors (NAI). He has been responsible for numerous research contracts totaling more than $20 million. He has authored over 700 publications and holds more than 30 patents, making significant contributions to the field of coal preparation, including coal flotation and washability analysis using X-ray computed tomography. His research on coal processing has been recognized with the Stefanko Awards and the Aplan Award from SME/AIME. He received the Lifetime Achievement Award at the XXIX International Mineral Processing Congress (IMPC) and the 2018 P.R. China International Science and Technology Cooperation Award in Beijing, China, in recognition of his outstanding contributions to mineral processing and international scientific collaboration.

Preface

This Special Issue, "Industrial Mineral Flotation—Fundamentals and Applications", addresses the growing scientific and industrial interest in flotation as a key separation process for industrial and critical minerals. The scope of this collection spans both fundamental studies—covering mineral–reagent interactions, interfacial chemistry, and bubble dynamics—and applied advances, such as rare-earth recovery, slag treatment, ultrasonic-assisted flotation, and emulsion flotation of fine coals. The aim is to provide readers with an integrated perspective on current challenges and future opportunities in flotation research and practice. The motivation for compiling this Special Issue arises from the increasing demand for sustainable mineral utilization and the role of flotation in resource efficiency and the circular economy. This work is intended for researchers, graduate students, and practitioners in mineral processing, materials science, and mining engineering. We gratefully acknowledge the contributions of all authors, the careful reviews provided by colleagues, and the support of the *Minerals* editorial team.

Xuming Wang and Jan D. Miller
Guest Editors

minerals

MDPI

Editorial

Editorial for the Special Issue "Industrial Minerals Flotation—Fundamentals and Applications"

Xuming Wang * and Jan D. Miller

Department of Materials Science and Engineering, University of Utah, Salt Lake City, UT 84112, USA; jan.miller@utah.edu
* Correspondence: x.wang@utah.edu

Industrial minerals are generally considered to be nonmetallic mineral resources [1]. Most industrial minerals, including limestone, clays, sand, gravel, diatomite, kaolin, bentonite, silica, barite, talc, and gypsum, are used in the construction industry and/or as additives, such as kaolinite in the production of paper [2–7]. The industrial minerals are valued for their physical and chemical properties, which make them useful for many industrial applications. In agriculture, industrial minerals such as potash and phosphate are essential fertilizer ingredients [8]. Bauxite is the primary source of aluminum ore and is also used to make ceramics, cement, and abrasives [9]. In the new energy sector, the industrial minerals also play an important supporting role. Graphite and manganese serve as electrodes in many lithium batteries [10,11]. Spodumene and various types of lithium clays are important sources of lithium for lithium-ion batteries [12]. Rare-earth resources are some of the most critical resources for many industries, including renewable energy and defense industries. As with many metallic mineral resources, concentration processes are needed to purify and enrich industrial minerals before further preparation and utilization. Flotation is a crucial process for the separation and concentration of industrial minerals from ores.

This Special Issue of *Minerals*, "Industrial Minerals Flotation—Fundamentals and Applications", presents eight contributions that collectively advance our understanding of flotation chemistry and interfacial phenomena while also highlighting new technological developments and industrial applications.

1. Advances in Fundamental Understanding

Four papers in this collection provide new insights into mineral–reagent interactions, bubble dynamics, and interfacial phenomena which occur during the flotation process.

Zhu et al. (Contribution 1) studied the floatability of spodumene minerals of different colors (pink, white, and purple). The color variations are attributed to the presence of surface polyvalent metal ions, such as Fe^{3+} and Mn^{2+}. Flotation experiments revealed distinct differences in the floatability of these colored spodumene samples. The authors further examined the adsorption and activation effects of calcium ions on spodumene of varying colors. Adsorption capacity measurements indicated that Ca^{2+} adsorption was higher on pink spodumene surfaces, which contain fewer polyvalent metal ions. Through experimental characterization and quantum chemical calculations, the study demonstrated how calcium species enhance collector adsorption and flotation performance, with significant implications for lithium resource processing.

Chen, Wang, and Zhang (Contribution 2) investigated the selective inhibition of fluorapatite by sulfuric acid during dolomite flotation. The effects of H_2SO_4 on the wettability and flotation selectivity of apatite and dolomite were evaluated using microflotation experiments and contact angle measurements. Molecular dynamics simulations (MDSs) revealed

 https://doi.org/10.3390/min15101038

that H_2SO_4 interacts with Ca sites on both fluorapatite and dolomite surfaces, thereby hindering the interaction of NaOl (sodium oleate collector) with Ca^{2+} sites. However, SO_4^{2-} ions do not prevent oleate adsorption on Mg sites of dolomite surfaces. In this way, H_2SO_4 selectively depresses fluorapatite while allowing dolomite to float.

Gungoren et al. (Contribution 3) examined the synergistic effects of common aqueous ions on flotation performance in Pb–Zn sulfide ores. Using artificial process waters, they studied dynamic surface tension, bubble size distribution, foam stability, and bubble coalescence. The results showed that increasing ionic strength reduced bubble size and narrowed the size distribution, while also improving foam stability. High concentrations of potassium ethyl xanthate (KEX) were effective in reducing surface tension. The study demonstrated that aqueous ions significantly influence bubble properties and flotation efficiency, emphasizing the importance of water chemistry in flotation practice.

Complementing these studies, Liu et al. (Contribution 4) provided a comprehensive review on the interfacial regulation of apatite flotation. They discussed the surface characteristics of apatite and its main gangue minerals, emphasizing the challenges posed by fine-grained ores. The review summarized advances in collector and depressant design, their mechanisms of action, and the regulation of wettability at the mineral–water interface. It also covered particle–particle interactions and particle–bubble attachment phenomena. The authors concluded by outlining prospects for future research, including the development of innovative reagents and improved interfacial control strategies for phosphate beneficiation.

Together, these works advance the molecular- to process-scale understanding of flotation, offering a strong basis for designing more selective and efficient reagent systems.

2. Innovative Techniques and Applications

The remaining four contributions emphasize technological innovation and applied solutions.

Wang et al. (Contribution 5) explored the recovery of xenotime and florencite from silicate minerals using a combination of magnetic separation and flotation. Initial magnetic separation enriched the rare earth minerals, raising the concentrate grade to 14.29% with 94.48% recovery. Subsequent flotation further improved the concentrate grade to 51.26% at a recovery of 90.47%. This combined approach achieved high efficiency in separating xenotime and florencite from quartz and illite gangue, demonstrating strong potential for industrial applications in rare earth processing.

Valderrama et al. (Contribution 6) investigated copper recovery from smelting slags at the Hernán Videla Lira Smelter in Chile. Laboratory flotation tests optimized grinding, pH, reagent schemes, and flotation times, producing concentrates with copper grades up to 24.4% and recoveries above 70%. Industrial-scale trials demonstrated even higher separation efficiency, with rougher–cleaner circuits yielding copper grades of 27.9% and recoveries of 87.5% while processing 1344 tons per day. This study highlights the feasibility of flotation for slag treatment, offering both environmental benefits and contributions to the circular economy.

Zhang et al. (Contribution 7) reviewed recent advances in ultrasonic-assisted flotation. They analyzed the effects of ultrasound on slurry properties, flotation processes, and device development, as well as results from pilot-scale testing. Ultrasonic pretreatment was shown to improve concentrate recovery by around 10%. The review summarized four key aspects: optimization of ultrasonic parameters, synergistic effects with reagents, effects on different particle size fractions, and applications to new mineral systems. The authors also discussed future prospects, including integration of ultrasonic flotation with magnetic fields, big data, sensors, and automated process control technologies.

Finally, Wang, Cheng, and Ding (Contribution 8) reviewed recent advances in emulsion flotation for difficult-to-float coals. These coals, characterized by high hydrophilicity and fine particle size, present significant challenges for conventional collectors. Emulsions, with strong interfacial regulation properties, have emerged as effective collectors, improving charge regulation, hydrophobicity, and interfacial tension. The review summarized the influence of emulsion type and preparation method on flotation efficiency, while also outlining future directions such as precise control of emulsion structure, development of low-cost and eco-friendly reagents, and improved storage and transportation methods.

These studies underscore how flotation continues to adapt to new challenges, from processing low-grade resources to recovering value from waste and advancing cleaner energy technologies.

3. Future Directions

The contributions in this Special Issue highlight two complementary trends in industrial mineral flotation research: (1) fundamental investigations that deepen mechanistic understanding, and (2) innovative engineering approaches that translate this knowledge into industrial practice. Together, they demonstrate that flotation remains a dynamic field, essential not only for traditional ores but also for industrial minerals, critical minerals, waste reprocessing, and extended use of coal resources.

Looking forward, the integration of computational modeling, novel reagent chemistry, high efficiency separation devices, and advanced process technologies with sustainable and digitalized mineral processing will be central to further progress. The convergence of flotation science with modeling, big data analytics, and green chemistry also represents an important direction for future progress.

We thank all the authors for their valuable contributions, the reviewers for their thoughtful evaluations, and the editorial team of *Minerals* for their support. We anticipate that these contributions will serve as a valuable reference for researchers and practitioners, while also inspiring new directions in the science and engineering of flotation.

Conflicts of Interest: The authors declare no conflicts of interest.

List of Contributions:

1. Zhu, G.; Wang, R.; Zheng, Y.; Zhang, X.; Zhang, Y.; Li, C.; Li, G.; Cao, Y. The Adsorption Characteristics of Calcium Ions on Spodumene with Different Colors and Their Associated Activation Mechanism. *Minerals* **2025**, *15*, 48. https://doi.org/10.3390/min15010048.
2. Chen, A.; Wang, X.; Zhang, Q. Interaction and Inhibition Mechanism of Sulfuric Acid with Fluorapatite (001) Surface and Dolomite (104) Surface: Flotation Experiments and Molecular Dynamics Simulations. *Minerals* **2023**, *13*, 1517. https://doi.org/10.3390/min13121517.
3. Gungoren, C.; Muse, S.M.; Terzi, M.; Eskibalci, M.F.; Unver, I.K.; Ozdemir, O. Synergistic Effect of Frequently Found Ions in the Flotation of Pb-Zn Sulfide Ores on Air/Water Interface. *Minerals* **2023**, *13*, 1236. https://doi.org/10.3390/min13101236.
4. Liu, Z.; Li, L.; Li, Z.; Wang, M.; Liu, F.; Mi, H. Review of Interfacial Regulation of Apatite Flotation. *Minerals* **2025**, *15*, 558. https://doi.org/10.3390/min15060558.
5. Wang, W.; Li, E.; Li, Z.; Zhu, W.; Wang, Y. Recovery of Xenotime and Florencite from Silicate Minerals Using a Combined Technique of Magnetic Separation and Flotation. *Minerals* **2024**, *14*, 1073. https://doi.org/10.3390/min14111073.
6. Valderrama, L.; Tapia, J.; Pavez, O.; Santander, M.; Rivera, V.; Gonzalez, M. Recovery of Copper from Slags Through Flotation at the Hernán Videla Lira Smelter. *Minerals* **2024**, *14*, 1228. https://doi.org/10.3390/min14121228.
7. Zhang, X.; Cheng, H.; Xu, K.; Ding, D.; Wang, X.; Wang, B.; Ma, Z. Ultrasonic Enhancement for Mineral Flotation: Technology, Device, and Engineering Applications. *Minerals* **2024**, *14*, 986. https://doi.org/10.3390/min14100986.

8. Wang, X.; Cheng, H.; Ding, D. Advances and Prospects on Flotation Enhancement of Difficult-to-Float Coal by Emulsion: A Review. *Minerals* **2024**, *14*, 952. https://doi.org/10.3390/min14090952

References

1. Dill, H.G. Kaolin: Soil, rock and ore: From the mineral to the magmatic, sedimentary and metamorphic environments. *Earth-Sci. Rev.* **2016**, *161*, 16–129. [CrossRef]
2. Singh, N.B. Clays and Clay Minerals in the Construction Industry. *Minerals* **2022**, *12*, 301. [CrossRef]
3. Murray, H.H. Traditional and new applications for kaolin, smectite, and palygorskite: A general overview. *Appl. Clay Sci.* **2000**, *17*, 207–221. [CrossRef]
4. Appiah, E.S.; Dzikunu, P.; Akinwamide, S.O.; Fangnon, E.A.K.; Mensah-Darkwa, K.; Andrews, A.; Agyemang, F.O.; Nartry, M.A.; Makgopa, K.; Bossuyt, S. A review on progress and prospects of diatomaceous earth as a bio-template material for electrochemical energy storage: Synthesis, characterization, and applications. *Ionics* **2024**, *30*, 7809–7860. [CrossRef]
5. Kipsanai, J.J.; Wambua, P.M.; Namango, S.S.; Amziane, S. A review on the incorporation of diatomaceous earth in construction materials. *Materials* **2022**, *15*, 7130. [CrossRef]
6. Wang, D.; Shi, C.; Farzadnia, N.; Shi, Z.; Jia, N.; Ou, Z. A review on the use of limestone powder in cement-based materials. *Constr. Build. Mater.* **2018**, *181*, 659–672. [CrossRef]
7. Esan, M.T. Review of gypsum-reinforced composites as building materials. *Discov. Civ. Eng.* **2024**, *1*, 5. [CrossRef]
8. Yahaya, S.M.; Mahmud, A.A.; Abdullahi, M.; Aruna, H.A. Recent advances in the chemistry of nitrogen, phosphorus and potassium as fertilizers in soil: A review. *Pedosphere* **2023**, *33*, 385–406. [CrossRef]
9. Kim, L.; Dobra, G.; Isopescu, R.; Iliev, S.; Cotet, L.; Boiangiu, A.; Filipescu, L. Short review on mechanical activation in non-metallurgical alumina production. *Rom. J. Ecol. Environ. Chem.* **2022**, *4*, 62–79. [CrossRef]
10. Zhao, W.; Zhao, C.; Wu, H.; Li, L.; Zhanga, C. Progress, challenge and perspective of graphite-based anode materials for lithium batteries: A review. *J. Energy Storage* **2024**, *81*, 110409. [CrossRef]
11. Vissers, D.R.; Chen, Z.; Shao, Y.; Engelhard, M.H.; Das, U.; Redfern, P.C.; Curtiss, L.A.; Pan, B. Role of Manganese Deposition on Graphite in the Capacity Fading of Lithium Ion Batteries. *ACS Appl. Mater. Interfaces* **2016**, *8*, 14244–14251. [CrossRef] [PubMed]
12. Liu, Y.; Ma, B.; Lü, Y.; Wang, C.; Chen, Y. A review of lithium extraction from natural resources. *Int. J. Miner. Metall. Mater.* **2023**, *30*, 209–224. [CrossRef] [PubMed Central]

minerals

Article

The Adsorption Characteristics of Calcium Ions on Spodumene with Different Colors and Their Associated Activation Mechanism

Guangli Zhu [1], Ruping Wang [1], Yan Zheng [1,2], Xu Zhang [1,2], Yuzhe Zhang [1,2], Chao Li [1], Guosheng Li [1,*] and Yijun Cao [1,2,3,*]

1 Zhongyuan Critical Metals Laboratory, School of Materials Science and Engineering, Zhengzhou University, No. 100 Science Avenue, Zhengzhou 450001, China; guanglizhu@zzu.edu.cn (G.Z.); wrp0529@163.com (R.W.); zhengyan19981207@163.com (Y.Z.); zx17344713352@163.com (X.Z.); zye081310@163.com (Y.Z.); c.li@zzu.edu.cn (C.L.)

2 School of Chemical Engineering, Zhengzhou University, No. 100 Science Avenue, Zhengzhou 450001, China

3 School of Materials Science and Engineering, Henan University of Science and Technology, No. 263 Kaiyuan Avenue, Luoyang 471023, China

* Correspondence: lgscumt@163.com (G.L.); yijuncao@126.com (Y.C.)

Abstract: This study investigates the activation behavior and mechanism of calcium ions on the flotation of spodumene with different colors. Using NaOL as a collector, in descending order of flotation recoveries were purple, pink, and white spodumene, while in the presence of $CaCl_2$, the flotation recoveries were increased and the order was pink, white, and purple spodumene. The zeta potential, adsorption amount, contact angle, and AFM measurements demonstrated that calcium ions adsorbed on the spodumene surface and promoted NaOL adsorption in alkaline conditions. Species distribution analysis showed that Ca^{2+}, $Ca(OH)^+$, and $Ca(OH)_2$ were essential components that play an activation role at pH = 12. The adsorption capacity and XPS results illustrated that $CaCl_2$ activated spodumene flotation in two ways. One possibility involves calcium ions and their hydroxyl compounds being adsorbed on the spodumene surface, where Ca and Al sites favored OL^- adsorption. The other possible way involves calcium ions and their hydroxyl compounds forming complexes with NaOL firstly in solution and then co-adsorbing on the spodumene surface. Quantum chemical calculations showed that the adsorption affinity on the spodumene surface in descending order was Ca^{2+}, $Ca(OH)_2$, and $Ca(OH)^+$, and the pink spodumene was most preferably adsorbed, followed by the white spodumene, also consistent the flotation results.

Keywords: flotation; spodumene with different colors; calcium ions; adsorption; activation mechanism

1. Introduction

Lithium is a strategically important metal that is extensively utilized in batteries, ceramics and glass, lubricants, and the nuclear and optoelectronic industries. With the rapid development of the lithium battery industry, there is rapid growth in lithium demand and consumption, and research suggests that lithium demand will see an average annual growth of 10% in the next ten years. Therefore, the safe supply of lithium is particularly important [1–6].

Spodumene is one of the main sources of lithium extraction [7]. During the flotation process, activators are commonly added to enhance the flotation efficiency of spodumene,

 https://doi.org/10.3390/min15010048

thereby enhancing the flotation separation of spodumene from its associated gangue minerals such as feldspar, albite, quartz, lepidolite, and muscovite [8,9]. The calcium ion is a vital activator commonly used in spodumene flotation. The effects of calcium ions on spodumene flotation and its activation mechanism have been studied by many researchers. Filippov et al. [3] observed that the flotation recovery of spodumene improved to about 90% with the addition of $CaCl_2$. Luo et al. [10] found that the primary active component that enhances the flotation separation of spodumene and albite is $Ca(OH)^+$, whereas $Ca(OH)_2$ is not favorable for flotation separation. Meng et al. [11] analyzed the effects of Fe^{3+}, Mg^{2+}, and Ca^{2+} on the flotation behavior of spodumene using NaOL as a collector. Their results showed that Fe^{3+}, Mg^{2+}, and Ca^{2+} had notable activation effects on spodumene at pH 6.5, 10.9, and 12.1, respectively, with Mg^{2+} and Ca^{2+} exhibiting better activation for spodumene than Fe^{3+}. Zhang et al. [12] indicated that the poor adsorption of Ca^{2+} makes it difficult to activate spodumene at a neutral pH, and $Ca(OH)^+$ effectively improved the adsorption strength of the collector, while $Ca(OH)_2$ greatly inhibited the collector's adsorption on Al sites. The species in pulp and adsorption form on the mineral surface are difficult to identify due to the limitations of detection means, and the mechanism by which calcium ions activate spodumene flotation remains controversial.

Spodumene has many colors, such as white, pink, purple, and green [13,14]. These colors are attributed to crystal imperfection, and some trace elements in particular, namely Fe, Mn, and Cr, are considered to be responsible for these colors. Rehman [15] and Rossman [16] found that spodumene contains many trace elements, while Fe and Mn are the major elements responsible for the differences in color. Ito et al. [17] observed that lilac coloration occurs when the Fe/Mn ratio is below 1. The Fe/Mn ratios were 0.35 for purple spodumene, 0.72 for pink spodumene, and 2.84 for white spodumene. The color of the mineral shifts from purple to pink and then to white with an increasing Fe/Mn ratio. There are differences in the flotability of spodumene with different colors in the same ore body. Sun [14,18] investigated the flotability of spodumene with three different colors, and the results indicated that the ratios of polyvalent metal cations to anions are the main reason for the difference in their natural flotability. Using sodium oleate as a collector, white spodumene had the best natural flotability, light green spodumene had the next best, and pink spodumene had the worst due to the fact that the white spodumene surface has the highest Fe^{3+} and Al^{3+} while the light green spodumene has lower O^{2-} and higher Fe^{3+} than pink spodumene. Fe^{3+} had the strongest activation effect on pink spodumene, followed by light green spodumene and white spodumene, and this is because the pink spodumene surface has more O^{2-} and is the most electronegative. The results indicate that Fe^{3+} shows different activation behaviors on the surface of spodumene with different colors, and also the trace metal elements on the mineral surface have a significant impact on mineral flotability.

In recent years, many studies have explored the adsorption behaviors of metal cations such as Fe^{3+}, Mg^{2+}, and Ca^{2+} on the spodumene surface and the influence mechanism on spodumene flotation, but there are few studies on the adsorption characteristics of metal cations on the surface of spodumene with different colors as well as their effects on flotation. Figuring out the activation mechanism of metal cations in different colored spodumene could enrich and develop the theory and technology of spodumene flotation. In this study, sodium oleate (NaOL), often employed in the industrial flotation of spodumene, was chosen as the collector [8,10,11,19], and the effects of calcium ions on the flotation of spodumene with white, pink, and purple were investigated using microflotation tests. The adsorption amount measurements, zeta potentials, contact angle, atomic force microscopy (AFM), X-ray photoelectron spectroscopy (XPS), and quantum chemical calculations were carried out to investigate the adsorption properties of calcium ions on the three spodumene surfaces as well as the influence mechanism of calcium ions on flotation.

2. Materials and Methods

2.1. Materials and Reagents

Spodumene crystals with white, pink, and purple colors were acquired from Western Sahara, Morocco. Spodumene samples of high purity were prepared through careful selection, crushing, grinding, and sieving. As the best flotation particle size in the spodumene flotation practice is −74 + 38 μm [8,9], −74 + 38 μm size fraction samples were used for microflotation tests. The mineralogical composition and chemical composition of white, pink, and purple spodumene were analyzed using X-ray diffraction (XRD) (Figure 1) and X-ray fluorescence (XRF) (Table 1), respectively. The XRD results show that the diffraction peaks for spodumene of three colors match those of the standard card of spodumene. The samples are mainly composed of Li_2O, SiO_2, and Al_2O_3, and the purity reaches more than 98%, which indicates that the samples are of high purity and can meet the requirements for single-mineral specimens. The XRF results show that the Fe/Mn ratio of white spodumene is greater than 1, that of pink spodumene is closer to 1, and that of purple spodumene is less than 1, and this was consistent with the findings of Ito et al. [17]. Fe and Mn are the major impurities responsible for color differences, as reported previously [20,21].

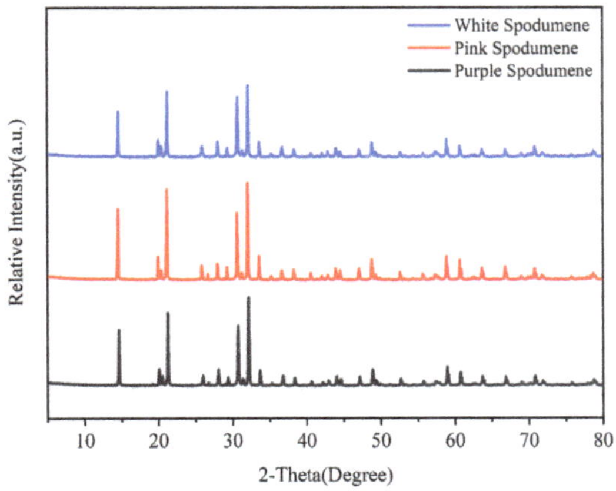

Figure 1. XRD patterns of spodumene samples with three colors.

Table 1. Chemical composition of spodumene samples with three colors (%).

Chemical Element	Li_2O	SiO_2	Al_2O_3	Na_2O	MnO	CaO	K_2O	P_2O_5	MgO	Fe_2O_3	TiO_2	Cr
White spodumene	7.62	63.81	26.67	0.804	0.165	0.318	0.074	0.060	0.047	0.468	0.013	0.0005
Pink spodumene	7.37	63.78	26.48	1.174	0.180	0.484	0.133	0.081	0.247	0.130	0.005	0.0003
Purple spodumene	7.43	63.58	26.86	1.140	0.179	0.609	0.122	0.067	0.033	0.062	0.014	0.0002

The experiments utilized deionized water that had a specific resistance of 18.3 MΩ·cm. Sodium hydroxide (96 wt%, Shangyang Palace, Luoyang, China) and hydrochloric acid (36 wt%, Shangyang Palace) were used as pH regulators in flotation. Sodium oleate (98 wt%, Macklin, Shanghai, China) was applied as the collector. All of the chemicals were analytical grade.

2.2. Microflotation Tests

The microflotation tests were carried out using XFG II flotation equipment (Wuhan Prospecting Machinery Factory, Wuhan, China), with a flotation cell of 40 mL and a stirring rate of 1600 rpm. For every experiment, 2 g of the mineral and 38 mL of deionized water were placed in the flotation cell. After stirring for 2 min, the pulp pH was adjusted with HCl or NaOH and then the system was stabilized for 3 min. $CaCl_2$ and NaOL were added sequentially and stirred for 2 min, respectively. Finally, the froth collection lasted for 5 min to obtain concentrates and tailings. The flotation products were filtrated, dried, and weighed separately, and flotation recoveries were calculated. Each condition was replicated three times and the results were averaged.

2.3. Zeta Potential Measurements

The zeta potential measurements were conducted using a Zetasizer Nano ZS90 Zeta potentiometer (Malvern Instruments, London, UK). The −38 μm size fraction was comminuted to −5 μm with an agate mortar for zeta potential measurements. Next, 20 mg samples and 38 mL of deionized water were added to a beaker and stirred for 2 min and then HCl or NaOH solution was added to adjust the slurry pH, followed by being stirred for another 5 min. Subsequently, $CaCl_2$ and NaOL were added in turn and mixed for 5 min. The system then rested for 60 min. Finally, the supernatant was collected for zeta potential measurements. The results for each sample were obtained by conducting three measurements and then calculating the average.

2.4. The Adsorption Amount Measurements

This measurement was carried out with an Avio-500 Inductively Coupled Plasma Mass Spectrometer (ICP-MS) (PerkinElmer Company, Shelton, CT, USA). First, 1 g samples and an appropriate amount of deionized water were poured into a beaker and stirred for 2 min. The slurry pH was regulated with HCl or NaOH, and after being stirred for 2 min, an appropriate amount of $CaCl_2$ was added, followed by being stirred for 40 min. Subsequently, the system was filtrated, and the filtrate was centrifuged with a high-speed centrifuge for 15 min at 9000 rpm. Then, the supernatant was taken and analyzed using ICP-MS to detect the metal ion content.

The amount of adsorption was measured as follows:

$$q_{eq} = \frac{(C_0 - C_{eq}) \times V}{m} \quad (1)$$

$$PAP = \frac{C_0 - C_{eq}}{C_0} \times 100\% \quad (2)$$

The Langmuir and Freundlich models were utilized to fit the nonlinear adsorption curves, aiming to better study the relationship between calcium ion adsorption capacity (q_{eq}) and its equilibrium concentration (C_{eq}). Langmuir isotherm models assume that the solute interacts chemically or physically with the adsorption sites on the surface of the adsorbent. The Freundlich adsorption isotherm is an empirical model, which is based on the assumption that the adsorption sites on the adsorbent surface are heterogeneous and that adsorption is multilayered [22]. The expression is as follows:

$$q_{eq} = \frac{q_{max} b C_{eq}}{1 + b C_{eq}} \quad (3)$$

$$q_{eq} = K_F C_{eq}^{1/n} \quad (4)$$

where q_{eq} is the proportion of metal ions adsorbed by spodumene per unit mass at the adsorption equilibrium (mg/g); C_0 is the primary concentration (mg/L); C_{eq} is the balanced concentration (mg/L); V is the slurry volume (L); m is the weight (g) of spodumene; PAP is the precipitation adsorption percentage (%); b is the adsorption constant; q_{max} is the maximum adsorption capacity of metal ions (mg/g); and K_F and n are the Freundlich empirical constants that characterize the adsorption capacity and adsorption strength, respectively.

2.5. Contact Angle Measurement

This measurement was carried out using a Theta Flow contact angle measuring instrument (Biolin, Goteborg, Sweden). The spodumene crystals of three colors were cut with a slicing machine to expose new surfaces, and then the crystals were stabilized in a block of epoxy resin with the surface exposed. The crystal surfaces were polished on an AutoMet 250Pro automatic grinding and polishing machine (Chicago, IL, USA) and then washed with deionized water and sequentially immersed in a fixed concentration of metal ions and/or NaOL solution for 40 min. Subsequently, the crystals were rinsed with deionized water and dried with nitrogen gas and then underwent contact angle measurements.

2.6. Atomic Force Microscopy (AFM) Imaging

The spodumene surface morphology in the presence of Ca^{2+} or NaOL was observed using the AFM-modeled JPK Nanowizard4 (JPK company, Berlin, Germany). The sample was prepared using the same procedure for the contact angle measurements.

2.7. X-Ray Photoelectron Spectroscopy (XPS) Analysis

The test was carried out using AXIS SUPRA XPS (Shimadzu Corporation, Kyoto, Japan). First, 0.5 g samples and an appropriate amount of deionized water were added to a beaker. After being stirred for 2 min, the suspension pH was stabilized using HCl or NaOH. After being stirred for 2 min, appropriate amounts of $CaCl_2$ and NaOL solution were added sequentially, and the solution was mixed for 3 min each. Then, the suspension was filtered, and the solid samples were vacuum dried at 60 °C for 24 h.

2.8. Quantum Chemical Simulation

Quantum chemical calculations were performed using Materials Studio 2018 founded on the density functional theory (DFT). The spodumene unit cell model was constructed using lattice parameters obtained from the American Mineralogist Crystal Structure Database. The (110) face was cut from the unit cell in the CASTEP module and the $2 \times 1 \times 1$ supercell model was created. Based on previous studies [20,21], Fe and Mn are the major impurities for the coloring of spodumene. Chemical composition analysis showed that the Fe/Mn ratio is greater than 1 in the white spodumene, while the Fe/Mn ratio is less than 1 in the purple spodumene and close to 1 in the pink spodumene. This conclusion is consistent with the literature [20,23–25]. The two metal cation sites M1 and M2 in the unit cell are aluminum ions and lithium ions, respectively [26,27]. The M1 position can be replaced by the transition metals Mn, Fe, Cr, etc., and M2 is mainly replaced by Na and K [7,28]. Thus, based on the chemical composition analysis, for white spodumene, the M1 position is replaced by two Fe^{3+} and one Mn^{2+}; for pink spodumene, the M1 position is replaced by one Fe^{3+} and one Mn^{2+}; and for purple spodumene, the M1 position is replaced by two Mn^{2+} and one Fe^{2+}. The models of the spodumene surfaces of three colors are shown in Figure 2.

The geometrical optimization and surface energy of the three spodumene samples were calculated using the CASTEP model. The DMol3 module played a role in optimizing the molecular structure of the calcium hydroxides and calculating the energy of the molecules. The adsorption configurations of Ca^{2+}, $Ca(OH)^+$, and $Ca(OH)_2$ on the surface of spodumene of three colors were constructed and the adsorption energies were calculated.

The adsorption energy was calculated as follows [29,30]:

$$E_{ads} = E_{surface+adsorbate} - (E_{surface} + E_{adsorbate}) \quad (5)$$

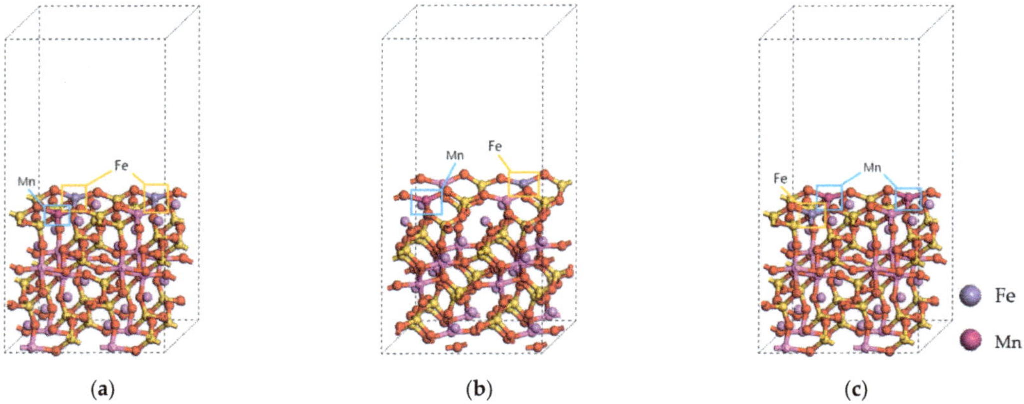

Figure 2. Surface models of spodumene of three colors: (**a**) white spodumene; (**b**) pink spodumene; (**c**) purple spodumene.

3. Results

3.1. Microflotation Tests

The microflotation test was conducted using 200 mg/L NaOL as the collector, in the absence of an activator, and the effects of pH in the flotation recoveries of white, pink, and purple spodumene were studied, as shown in Figure 3. As can be seen in Figure 3, the recoveries all increase as the pH rises and peak at about pH = 7 and decrease with the pH rising further. However, it is obvious that there is a difference in the flotation behaviors of the three spodumene samples, and the flotation recoveries in descending order are purple spodumene, pink spodumene, and white spodumene. White spodumene has the lowest flotation recovery rate. As reported [14], the oleate adsorption on the spodumene surface is mainly dominated by the Al sites on the surface. This could be one of the reasons for the flotation differences.

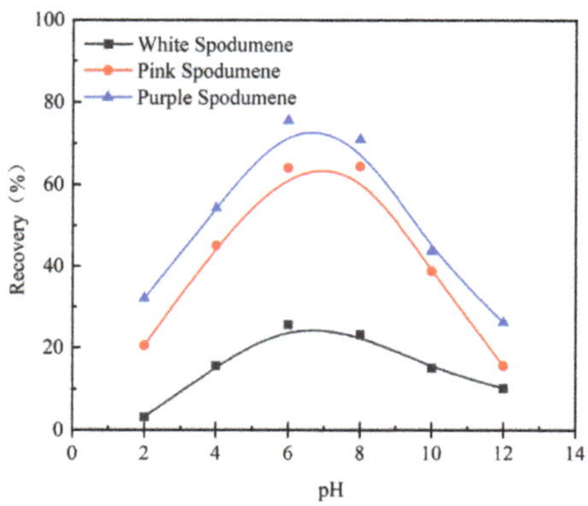

Figure 3. Recovery of the three spodumene samples as a function of pH with a dosage of 200 mg/L NaOL.

The flotation recoveries of spodumene of three colors as a function of pH in the presence of 100 mg/L $CaCl_2$ and 200 mg/L NaOL are shown in Figure 4. The flotation recoveries show an increasing trend with pH increasing and in descending order are pink, white, and purple spodumene. The recoveries are low when the pH is 6–11, whereas when the pH is increased to 12, the recoveries start to increase significantly, and the highest recovery is achieved at pH 13, with flotation recoveries of 35.97%, 59.06%, and 17.95% for white, pink, and purple spodumene, respectively. This shows that the optimal pH interval for Ca^{2+} activation is strong alkalinity. Considering industrial practice, the pulp pH is generally about 12, and therefore, at pH 12, the flotation recoveries of spodumene of three colors as a function of $CaCl_2$ dosage in the presence of 200 mg/L NaOL were studied further. As illustrated in Figure 5, $CaCl_2$ dosage has a significant role in spodumene flotation, and as the $CaCl_2$ dosage increases, the flotation recoveries for the three spodumene samples first increase and then decrease. The optimal $CaCl_2$ dosages for white, pink, and purple spodumene are 3.3×10^2 mg/L, 2.2×10^2 mg/L, and 2.2×10^2 mg/L, respectively. The highest recovery values of pink, white, and purple spodumene flotation are 74.19%, 72.33%, and 59.86%, respectively.

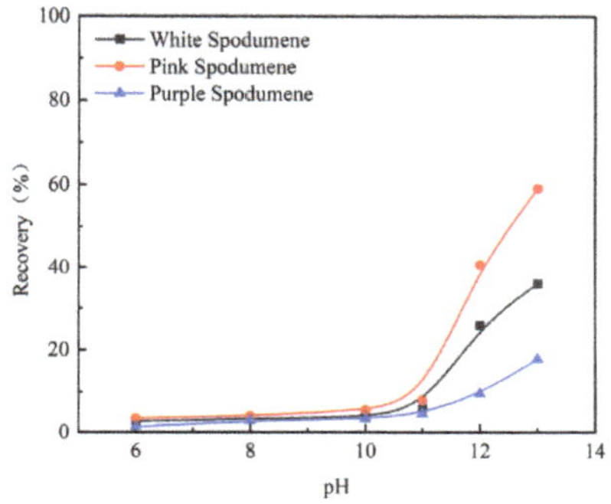

Figure 4. Recovery of the three spodumene samples as a function of pH with dosages of 100 mg/L $CaCl_2$ and 200 mg/L NaOL.

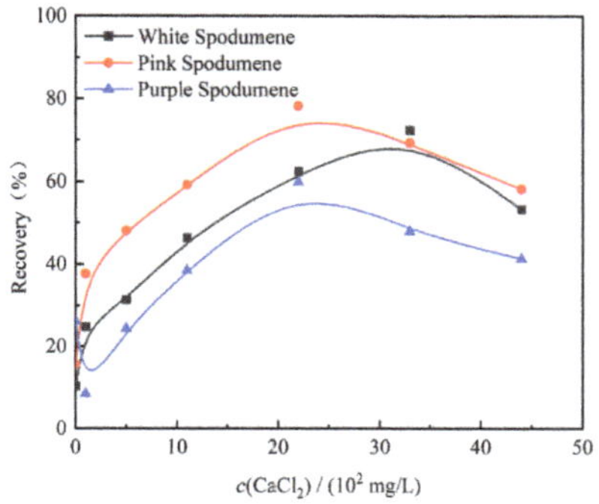

Figure 5. Recovery of the three spodumene samples as a function of $CaCl_2$ dosage with 200 mg/L NaOL at pH 12.

3.2. Zeta-Potential Analysis

Figure 6a shows, in the presence and absence of Ca(II) ions, the zeta-potentials of spodumene as a function of pH with or without NaOL. The points of zero charge (PZC) of white, pink, and purple pure spodumene occur at pH 2.10, 2.19, and 2.04, respectively. However, the PZC of spodumene is reported to be around 2.8 [31,32], and this may be due to the different origins of spodumene. The zeta potentials decrease with pH increasing and reach their most negative potential at pH 12, consistent with reports [33–36]. In the presence of Ca^{2+}, the zeta potentials move positively as a whole, indicating that Ca^{2+} is adsorbed on the negatively charged surface due to electrostatic attraction. The zeta potentials of spodumene exhibit an overall negative shift in the presence of NaOL. This is due to the adsorption of anionic OL^- on the spodumene surface, which decreases the surface positive charge, thus lowering the zeta potential.

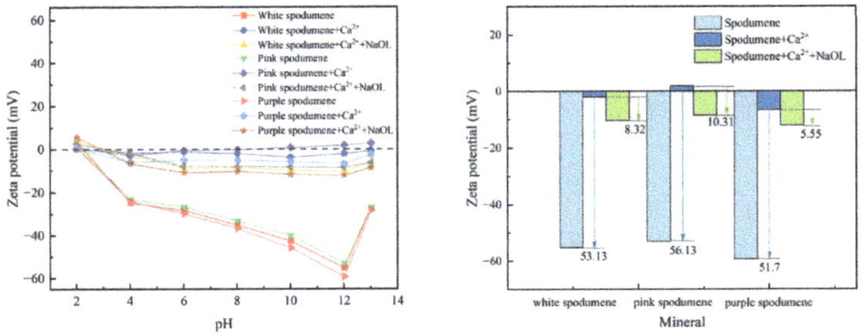

(a) (b)

Figure 6. (**a**) Zeta potentials of spodumene as a function of pH and (**b**) histogram of the changes in the potential of spodumene before and after the action of Ca^{2+} and NaOL at pH 12.

Figure 6b shows the difference in zeta potentials with and without NaOL and Ca^{2+} at pH 12. As can be seen, after the action of Ca^{2+}, the potential difference values of the three spodumene samples in descending order are pink, white, and purple spodumene. After NaOL addition, the zeta potentials for the three spodumene become more negative, and the potential difference values in descending order are pink, white, and purple spodumene, consistent with the flotation results.

3.3. Species Distribution Analysis

In solution, metal ions undergo hydrolysis, resulting in various hydroxyl complexes and hydroxide precipitates [37–39]. The concentrations of the hydrolysis component of calcium ions as a function of pH were calculated and the results are shown in Figure 7. From Figure 7a, it can be seen that for a solution of calcium ions with a density of 2×10^{-2} mol/L, Ca^{2+} is the main hydrolyzed component in the range of solution pH of 2–12.241. $Ca(OH)^+$ and $Ca(OH)_{2(aq)}$ begin to appear in the solution at pH > 4.3 and pH > 8.4, respectively, and the concentrations rise with the pH increasing. At pH > 12.241, the calcium hydroxide precipitates $Ca(OH)_{2(s)}$ form, with $Ca(OH)^+$ and $Ca(OH)_{2(aq)}$ as the main hydrolyzed components. For the calcium ion solution with a density of 3×10^{-2} mol/L, the species distribution is similar to that of $C_T = 2 \times 10^{-2}$ mol/L, and Ca^{2+}, $Ca(OH)^+$, and $Ca(OH)_{2(aq)}$ are the main components at pH 12.

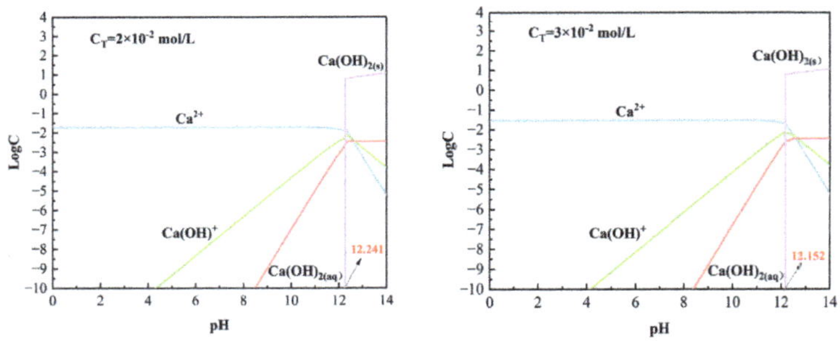

(a) (b)

Figure 7. Species distribution diagrams of Ca(II) in solution as a function pH: (**a**) $C_{(CaCl_2)} = 2 \times 10^{-2}$ mol/L and (**b**) $C_{(CaCl_2)} = 3 \times 10^{-2}$ mol/L.

3.4. Adsorption Capacity Measurements

Ca^{2+} adsorption on the spodumene surfaces as a function of $CaCl_2$ dosage at pH 12 is shown in Figure 8. The adsorption amount of Ca^{2+} increases with the $CaCl_2$ dosage increasing. The fitting curves and related parameters of Freundlich and Langmuir obtained through fitting analysis are shown in Figure 9 and Table 2. The Freundlich fit variance R_2 is generally higher than the Langmuir fit variance for the three spodumene samples, where R_2 represents the degree to which the model fits the experimental data. This indicates that the experimental results are in better anastomosis with the Freundlich adsorption isothermal model. Based on the Freundlich model assumption, the adsorption behavior of calcium ions on the spodumene surface at pH 12 is not limited to monolayer adsorption. K_F characterizes the adsorption capacity, and the K_F values of white, pink, and purple spodumene are 0.37, 0.78, and 0.32, respectively, indicating that calcium ions have a stronger adsorption ability on the pink spodumene surface. Based on this, it was hypothesized that the flotation of pink spodumene would be better than the white and purple spodumene, which is consistent with the flotation results.

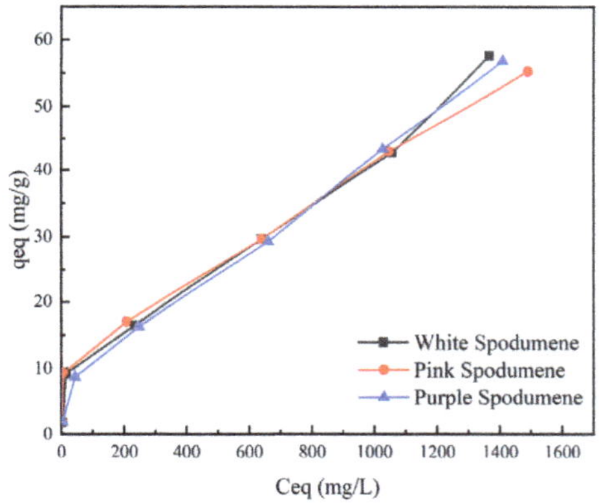

Figure 8. Adsorption curves of spodumene of different colors as a function of $CaCl_2$ dosage at pH 12.

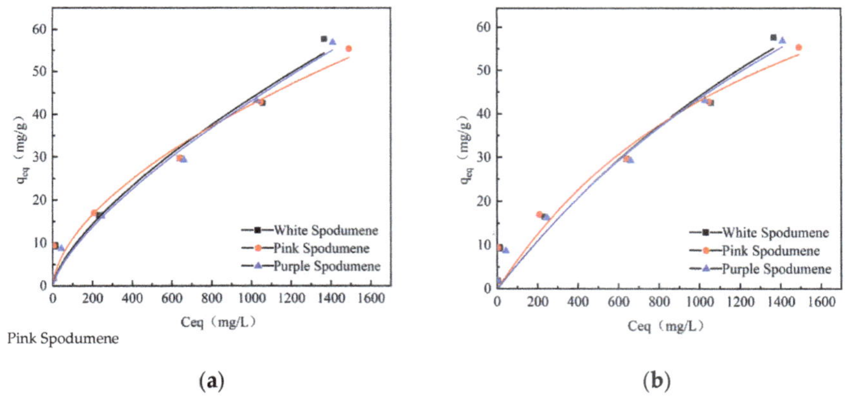

Figure 9. Fitting curves for Ca^{2+} adsorption of spodumene: (**a**) Freundlich and (**b**) Langmuir.

Table 2. The fitting parameters of Ca^{2+} adsorption isotherm curves on the three spodumene samples.

Colors	Langmuir			Freundlich		
	q_{max} (mg/g)	b (L/mg)	R_2	K_F	1/n	R_2
White	173.19	3.42	0.94	0.37	0.69	0.96
Pink	108.15	6.63	0.93	0.78	0.58	0.96
Purple	163.52	3.65	0.97	0.32	0.71	0.98

3.5. Contact Angle Analysis

The contact angles are illustrated in Figure 10. The hydrophobicity of the three spodumene samples in descending order is pink, white, and purple. After the addition of Ca^{2+}, the contact angle on the spodumene surface goes up. For white, pink, and purple spodumene, the average contact angles are 43.75°, 63.75°, and 48.40°, respectively. When treated with NaOL only, the average contact angles for white, pink, and purple spodumene surfaces are 48.27°, 54.43°, and 58.57°, respectively. After Ca^{2+} and NaOL treatment, the contact angles on the surface of all three spodumene samples reach more than 75°, and the surface contact angles for white, pink, and purple spodumene are 84.50°, 93.06°, and 74.15°, respectively. The contact angle on the surface of pink spodumene is larger, indicating that the pink spodumene has greater hydrophobicity, which is in agreement with the micro-flotation tests.

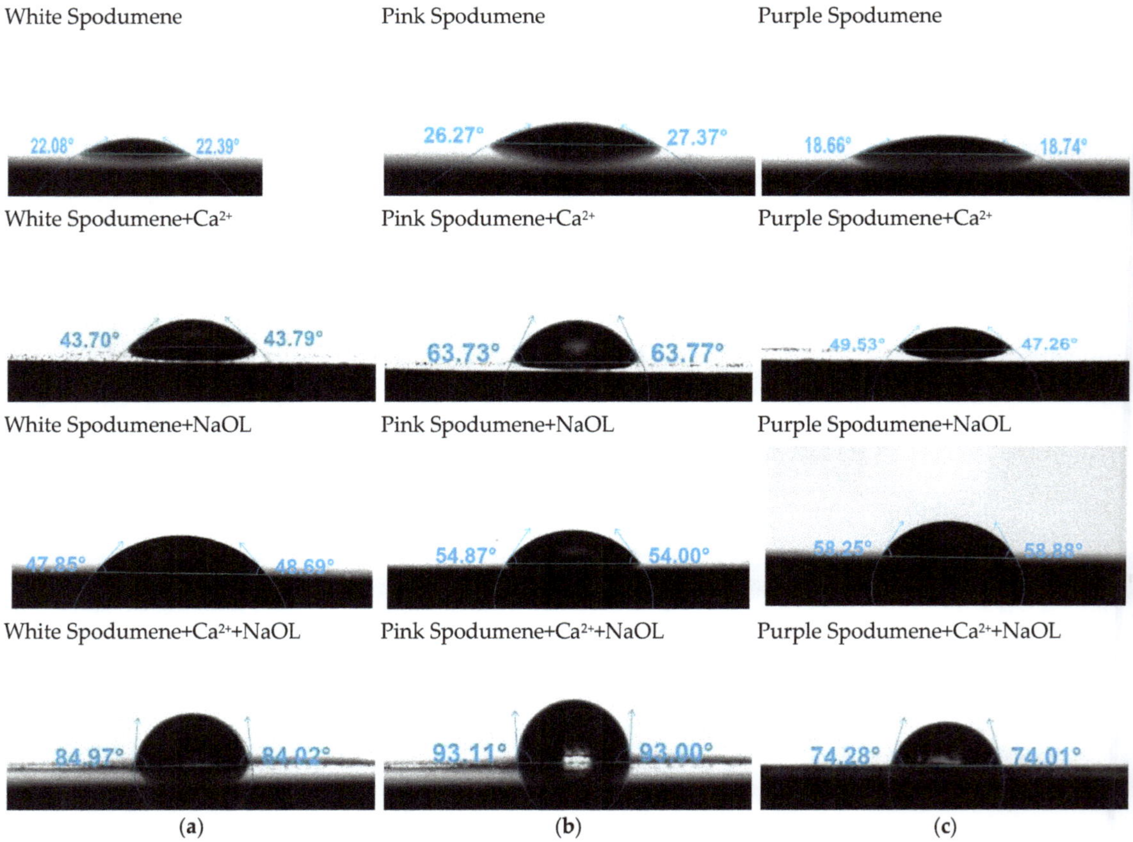

Figure 10. Contact angles of spodumene before and after reagent treatment: (**a**) white spodumene; (**b**) pink spodumene; (**c**) purple spodumene.

3.6. AFM Analysis

The surface morphology of the white, pink, and purple spodumene samples before and after the interaction with calcium ions and sodium oleate is shown in Figures 11a, 12a and 13a. The surface roughness of spodumene before and after the action of Ca^{2+} and NaOL is presented in Table 3. Before reagent addition, the maximum height and average roughness Ra on the surface of white, pink, and purple spodumene are 1.2 nm and 0.209 nm, 7.23 nm and 1.185 nm, and 1.57 nm and 0.251 nm, respectively, which indicates that the surfaces of the polished spodumene are relatively smooth, with only a few small bumpy grains. As shown in Figures 11b, 12b and 13b, in the presence of Ca^{2+}, the surfaces of spodumene are very rough, indicative of Ca^{2+} adsorption. After the addition of NaOL, as shown in Figures 11c, 12c and 13c, the surface roughness increases, and there are little dots that spread evenly on the spodumene surface, illustrating that NaOL is uniformly adsorbed on the surface. The increase in roughness indicated NaOL adsorption, and the maximum height and average roughness on the surface of white, pink, and purple spodumene after NaOL treatment reached 15.2 nm and 0.910 nm, 19.8 nm and 2.620 nm, 13.9 nm and 0.787 nm, respectively, which indicates that there was NaOL adsorption on the three spodumene surfaces and that there may be more NaOL adsorption on the surface of the pink spodumene, consistent with the flotation results.

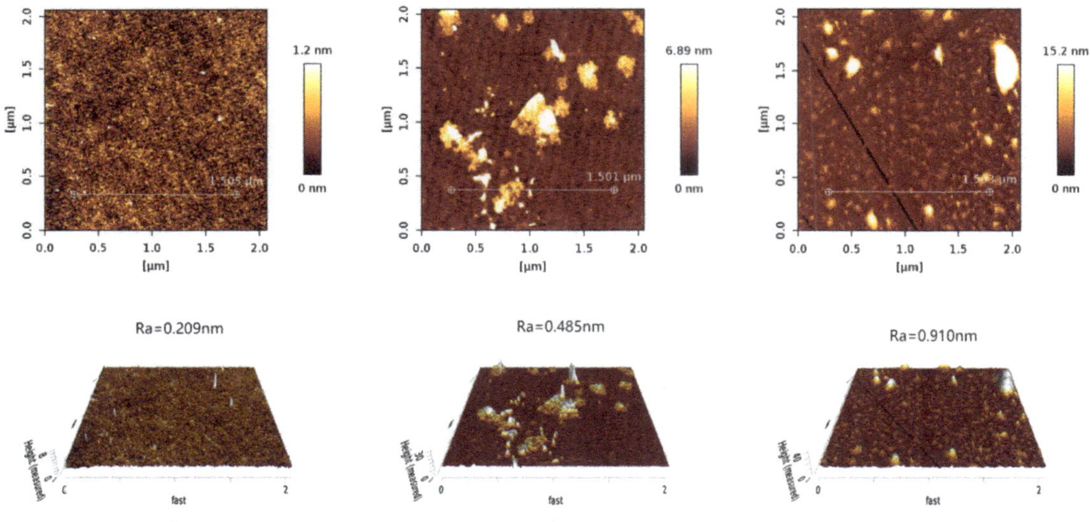

Figure 11. Two-dimensional and three-dimensional images of AFM before and after interaction between the white spodumene and the reagents: (**a**) white pure spodumene; (**b**) in the presence of Ca^{2+}; (**c**) in the presence of Ca^{2+} and NaOL.

Table 3. The surface roughness of spodumene before and after reagent treatment, as determined using AFM.

Minerals	Ra (nm)		
	Spodumene	Spodumene + Ca^{2+}	Spodumene + Ca^{2+} + NaOL
White spodumene	0.209	0.485	0.910
Pink spodumene	1.185	1.619	2.620
Purple spodumene	0.251	0.443	0.787

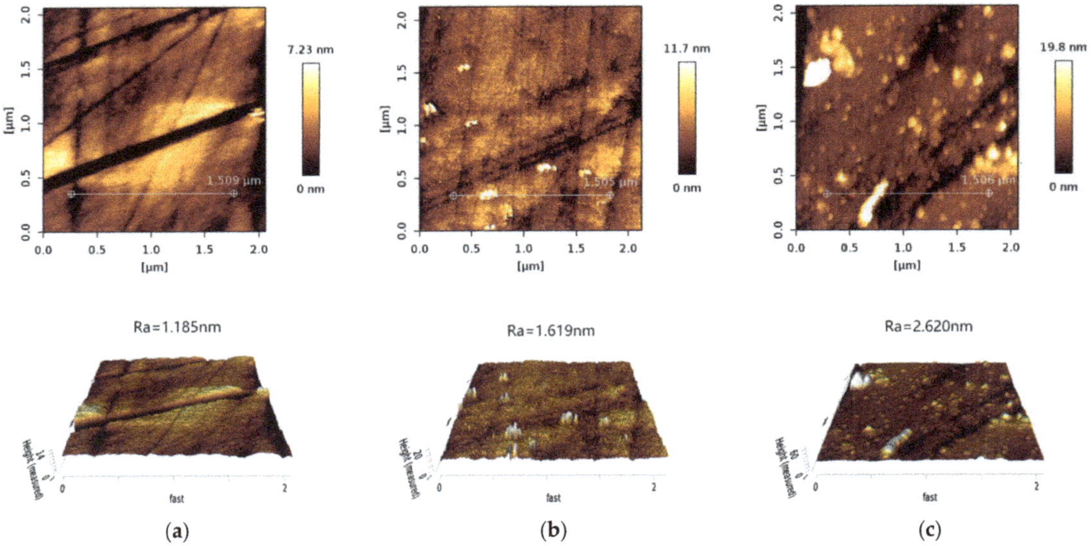

Figure 12. Two-dimensional and three-dimensional images of AFM before and after interaction between pink spodumene and the reagents, (**a**) pink pure spodumene; (**b**) in the presence of Ca^{2+}; (**c**) in the presence of Ca^{2+} and NaOL.

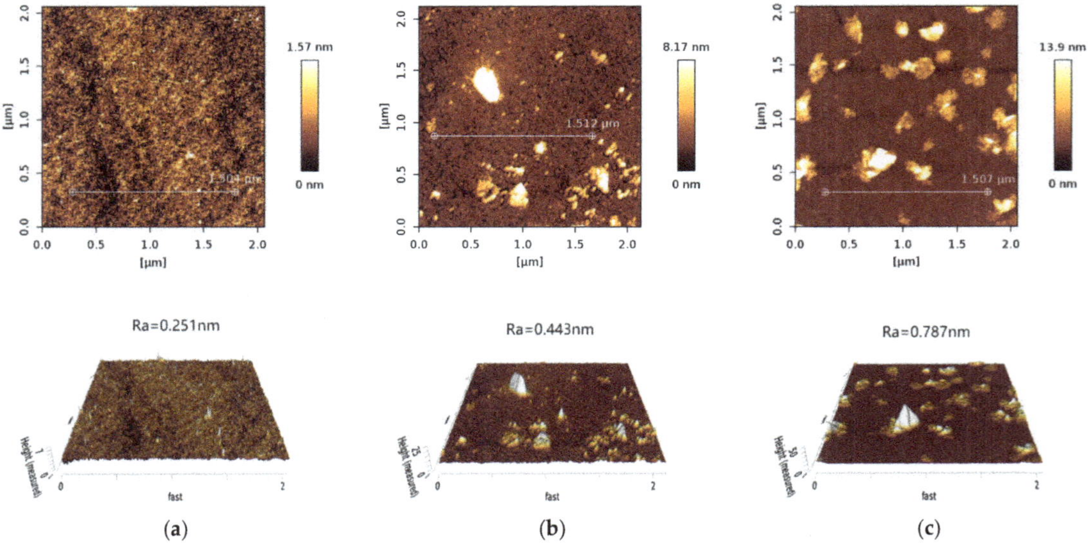

Figure 13. Two-dimensional and three-dimensional images of AFM before and after the interaction between the purple spodumene and the reagents: (**a**) purple pure spodumene; (**b**) in the presence of Ca^{2+}; (**c**) in the presence of Ca^{2+} and NaOL.

3.7. XPS and Calcium Ion Adsorption Analysis

Figures 14–17 display the XPS spectra of the spodumene surface with and without Ca^{2+} and NaOL, and the elemental content on the spodumene surface is shown in Table 4. The adsorption of calcium ions before and after NaOL adsorption was studied and the results are illustrated in Figure 18. Table 4 shows that the white spodumene surface has the lowest Al content, which may be the reason why white spodumene has poor flotation performance. Besides, the Ca content on the surface of the three spodumene samples increases after calcium ion addition, indicating that calcium ions are adsorbed on the surface. After NaOL addition, the C content on the surfaces of spodumene with three colors notably rose, indicating that NaOL adsorbed on the surface. However, the Ca content is also relatively elevated, as shown in Figure 18. This may be due to OL^- chemically reacting with calcium ions and/or the hydroxyl compounds in the solution and then the complex co-adsorbing on the spodumene surface.

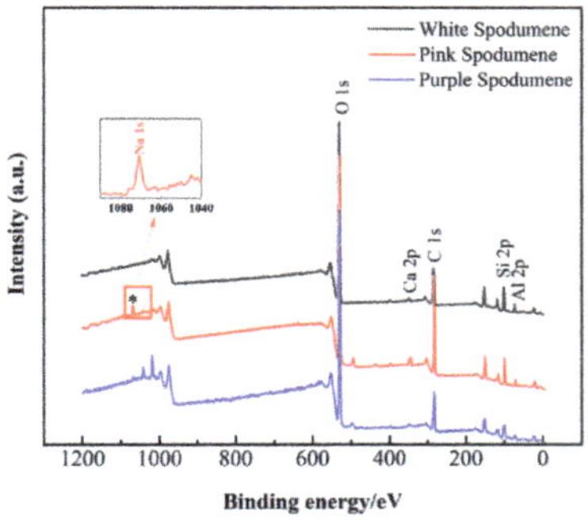

Figure 14. XPS full spectrum of spodumene of three colors.

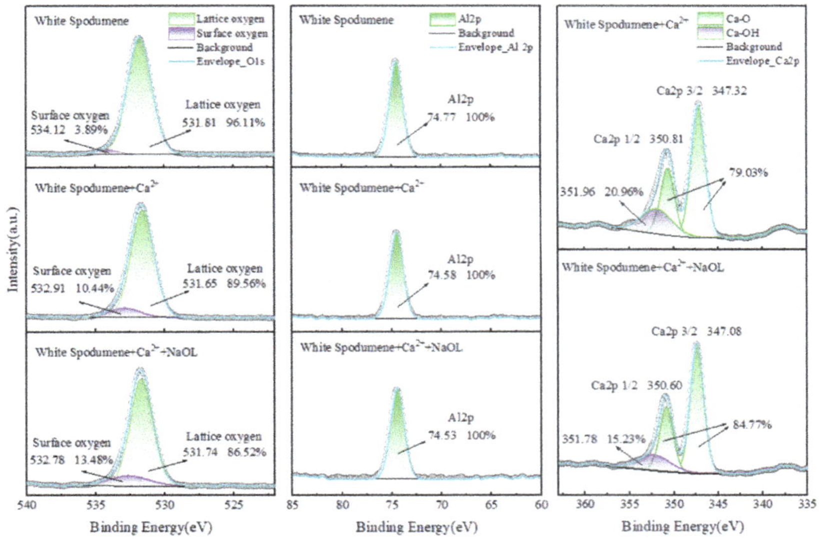

Figure 15. XPS spectra of the white spodumene surface before and after reagent adsorption.

Figure 14 shows the full spectra of the three spodumene samples. There is no peak for Fe and Mn, attributed to the low level of Fe and Mn. As depicted in Figures 15–17, the binding energy near 74.7 eV is attributed to Al 2p, the peaks near 534.1 eV and 531.7 eV are attributed to surface oxygen and lattice oxygen [40,41], and the peaks near 350.9 eV, 247.4 eV, and 351.9 eV are attributed to Ca 2p1/2, Ca 2p3/2, and Ca-OH, respectively. After the activation of Ca^{2+}, the peak for surface oxygen undergoes an obvious chemical shift and the integral area becomes larger in percentage, indicating that Ca has bonded with O elements on the spodumene surface. The peaks of surface oxygen of white, pink, and purple spodumene were negatively shifted by 0.21 eV, 0.22 eV, and 0.13 eV, respectively, and the chemical shift of the peaks of surface oxygen of pink spodumene was obvious, which indicated that the surface of pink spodumene was better activated by Ca^{2+}. The Ca 2p and Al 2p peaks were negatively shifted after NaOL treatment, suggesting that the chemical surroundings of Ca and Al on the spodumene surface have been altered and the chemisorption of NaOL at the Ca and Al sites occurs. The peaks at Al 2p, Ca 2p1/2, and Ca 2p3/2 of white, pink, and purple spodumene were negatively shifted by 0.05 eV, 0.21 eV, 0.24 eV; 0.11 eV, 0.36 eV, 0.41 eV; and 0.06 eV, 0.14 eV, 0.15 eV, respectively. The chemical shift of the separation peaks at Al 2p and Ca 2p of the pink spodumene surface is most significant, followed by white and purple spodumene, which is consistent with the flotation results.

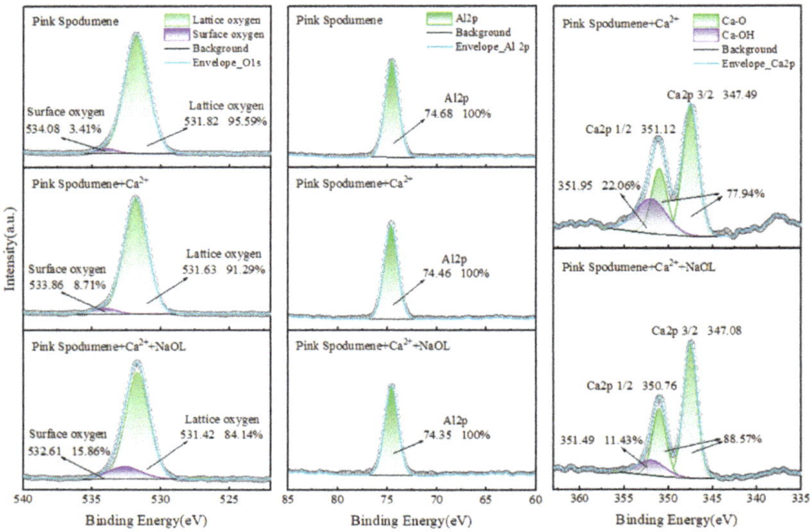

Figure 16. XPS spectra of the pink spodumene surface before and after reagent adsorption.

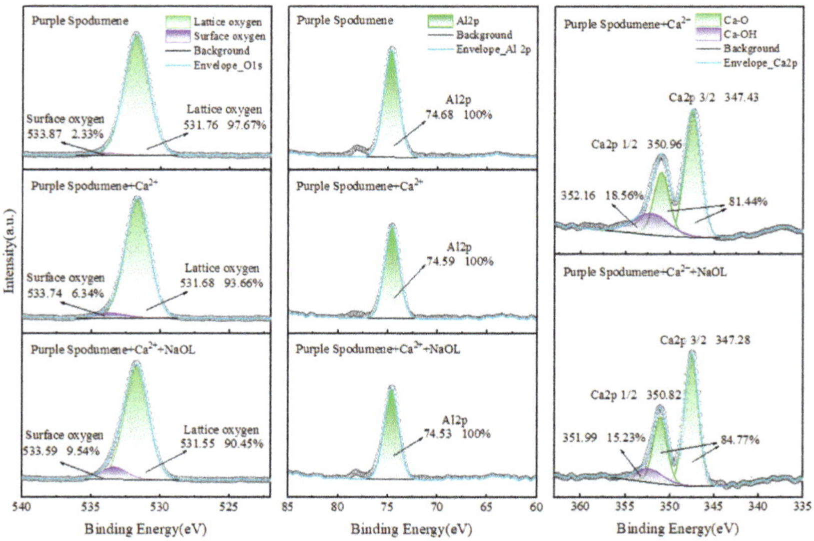

Figure 17. XPS spectra of the purple spodumene surface before and after reagent adsorption.

Table 4. Relative elemental content on the spodumene surface in the presence and absence of reagents.

Sample	Relative Content (%)				
	C	O	Ca	Si	Al
White spodumene	24.63	50.26	0.64	16.88	7.59
White spodumene + Ca^{2+}	28.68	45.82	0.72	16.20	8.58
White spodumene + Ca^{2+} + NaOL	40.08	38.39	0.97	13.54	7.01
Pink spodumene	33.98	42.19	0.42	15.51	7.91
Pink spodumene + Ca^{2+}	34.96	41.37	0.75	15.71	7.21
Pink spodumene + Ca^{2+} + NaOL	51.70	30.41	1.39	11.12	5.38
Purple spodumene	31.15	45.31	0.38	14.79	8.37
Purple spodumene + Ca^{2+}	32.29	43.81	0.88	15.12	7.90
Purple spodumene + Ca^{2+} + NaOL	55.65	28.43	1.11	9.42	5.39

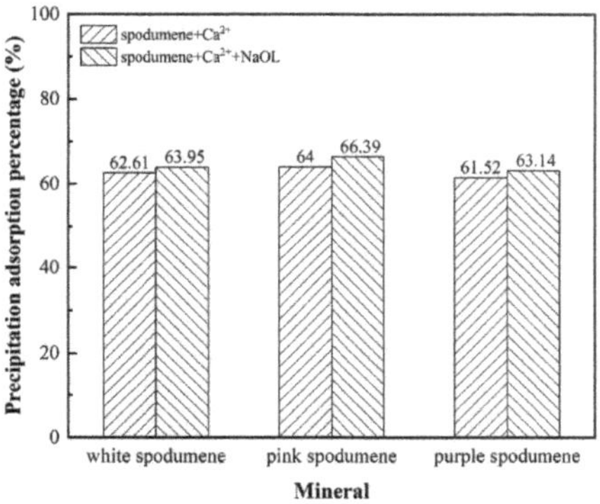

Figure 18. Histogram of the percentage of calcium ion adsorption and precipitation in the presence and absence of NaOL.

3.8. Quantum Chemical Simulation

The solution chemistry diagram of calcium ions shows that the major acting components of calcium ions in solution are Ca^{2+}, $Ca(OH)^{+}$, and $Ca(OH)_2$ at pH 12. The concentrations of the three components from high to low are: Ca^{2+}, $Ca(OH)^{+}$, and $Ca(OH)_2$. Therefore, the adsorption configurations of Ca^{2+}, $Ca(OH)^{+}$, and $Ca(OH)_2$ on the spodumene surface of three colors were constructed and the adsorption energies were calculated, as shown in Figure 19 and Table 5, respectively. Dashed lines represent the bond lengths between Ca atoms and O atoms in calcium ions and their hydroxide compounds, and O atoms and metal atoms on different colored diopside surfaces, all of which are indicated in the image. The numbers represent bond lengths, with the unit being Å The results show that the adsorption energies for calcium ions and their hydrolysis products on the surfaces of all three spodumene samples are negative, and the absolute values in descending order are Ca^{2+}, $Ca(OH)_2$, and $Ca(OH)^{+}$. The absolute values of adsorption energies for calcium ions and their hydrolysis compounds on the spodumene surface of three colors in descending order are pink, white, and purple spodumene, indicating that the adsorption of calcium ions and their hydrolysis compounds on the pink spodumene is better, followed by the white spodumene, which is in agreement with the microflotation results. As the pink spodumene has one Fe^{3+} and one Mn^{2+}, compared to the white spodumene (two Fe^{3+} and one Mn^{2+}) and purple spodumene (two Mn^{2+} and one Fe^{2+}), the pink spodumene surface has fewer polyvalent metal ions, which will favors the adsorption of calcium ions and their hydroxylated compounds. The white spodumene has the most polyvalent metal ions, not conducive to calcium adsorption.

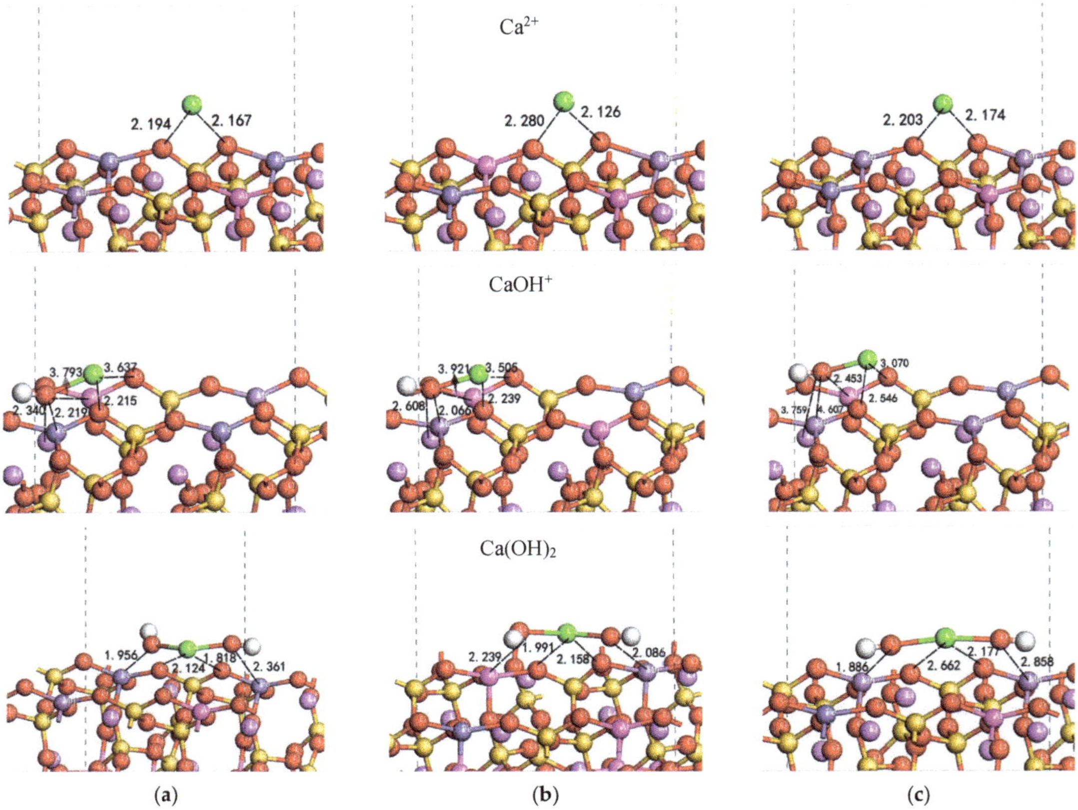

Figure 19. Adsorption configurations of calcium ions and their hydroxyl compounds on the surface of (**a**) white spodumene; (**b**) pink spodumene; and (**c**) purple spodumene.

Table 5. Adsorption energy of calcium ions and their hydroxyl compounds on the spodumene surfaces (eV).

Acting Components	Mineral Color	$E_{surface+adsorbate}$	$E_{adsorbate}$	$E_{surface}$	E_{ads}
Ca^{2+}	White	−24,113.77600	−676.79725	−23,436.70830	−7.360
	Pink	−23,092.71119	−676.79725	−22,415.63322	−7.639
	Purple	−24,001.04950	−676.79725	−23,323.99613	−6.969
$Ca(OH)^+$	White	−24,189.78006	−752.956724	−23,436.67971	−3.908
	Pink	−23,168.77055	−752.957945	−22,415.65643	−4.250
	Purple	−24,077.08097	−752.989768	−23,323.99575	−2.598
$Ca(OH)_2$	White	−24,265.84491	−829.02002	−23,436.67051	−4.201
	Pink	−23,244.79134	−829.01777	−22,415.60910	−4.476
	Purple	−24,153.13202	−829.03372	−23,323.99575	−2.791

4. Conclusions

Using NaOL as a collector, our flotation test results show that calcium ions have the best activation effect on the pink spodumene, followed by white and purple spodumene. The contact angle, zeta potential, AFM analysis, and adsorption capacity measurements all indicate more Ca species and NaOL adsorption on the surface of the pink spodumene, followed by white and purple spodumene; this corresponds to the flotation test outcomes. Spodumenes of different colors show different flotabilities, and the relative amounts of metal elements like Fe and Mn on their surfaces are key factors.

The XPS and adsorption capacity measurement results illustrate two activation patterns of calcium ions on the spodumene surface. One mode involves calcium ions and their hydroxyl compounds being adsorbed on the spodumene surface, which is favorable for OL^- adsorption, where chemical adsorption occurs between OL^- and the Ca and Al sites. The other mode involves calcium ions and their hydroxyl compounds forming colloidal complexes with NaOL firstly in solution and then co-adsorbing on the spodumene surface.

Species distribution analysis and quantum chemical calculations show that Ca^{2+}, $Ca(OH)^+$, and $Ca(OH)_2$ are the essential components that play an activation role at pH = 12, and the adsorption affinity on the spodumene surface in descending order was Ca^{2+}, $Ca(OH)_2$, and $Ca(OH)^+$. The adsorption of calcium ions and their hydrolysis compounds on the pink spodumene is better, followed by the white spodumene, which further confirmed the microflotation results.

The activation mechanism of calcium ions on the flotation of spodumene with different colors enriches theoretical research on spodumene flotation. As the metal ions activate spodumene with different colors to different extents, color should be considered in spodumene flotation practice. In addition, the varieties, contents, valence state, and positions within the crystal structure and on the surface and their effect on spodumene flotation have significant theoretical significance for the design and synthesis of spodumene flotation reagents.

Author Contributions: Conceptualization, G.Z.; methodology, R.W.; software, Y.Z. (Yan Zheng); validation, X.Z. and Y.Z. (Yuzhe Zhang); formal analysis, R.W.; investigation, R.W.; resources, Y.C.; data curation, C.L.; writing—original draft preparation, R.W.; writing—review and editing, G.Z.; visualization, X.Z.; supervision, Y.C.; project administration, G.L. and Y.C.; funding acquisition, G.Z., C.L. and Y.C. All authors have read and agreed to the published version of the manuscript.

Funding: This research was funded by the Project of Zhongyuan Critical Metals Laboratory (Grant No. GJJSGFYQ202320 and GJJSGFZD202304), the National Natural Science Foundation of China (Grant No. 52204287), the Zhengzhou Key Project for Collaborative Innovation (Grant No. 21XTZX06027), the Key R & D and Promotion Program in Henan Province (Grant No. 242102321176), and the National Supercomputing Center in Zhengzhou.

Data Availability Statement: Data are contained within the article.

Conflicts of Interest: The authors declare no conflicts of interest.

References

1. Liu, D.; Gao, X.; An, H.; Qi, Y.; Sun, X.; Wang, Z. Supply and demand response trends of lithium resources driven by the demand of emerging renewable energy technologies in China. *Resour. Conserv. Recycl.* **2019**, *145*, 311–321. [CrossRef]
2. Guo, Z.F.; Tan, C.G.; Dai, D. Lithium-the new focus of the world's mining development. *Chin. Nucl. Ind.* **2017**, *9*, 28–29.
3. Filippov, L.; Farrokhpay, S.; Lyo, L.; Filippova, I. Spodumene Flotation Mechanism. *Minerals* **2019**, *9*, 6. [CrossRef]
4. Burt, R.O.; Zhang, L. Denidae—A new type of low-iron spodumene. *Miner. Prot. Util.* **1989**, *5*, 52–55.
5. James, S.B.; Barrett, H.C.; Pan, C.L. Flotation separation of spodumene and beryl. *Min. Eng.* **1963**, *1*, 28–33.
6. Dessemond, C.; Lajoie-Leroux, F.; Soucy, G.; Laroche, N.; Magnan, J.F. Spodumene: The Lithium Market, Resources and Processes. *Minerals* **2019**, *9*, 6. [CrossRef]
7. Zhu, G.; Cao, Y.; Wang, Y.; Wang, X.; Miller, J.D.; Lu, D. Surface chemistry features of spodumene with isomorphous substitution. *Miner. Eng.* **2020**, *146*, 5. [CrossRef]
8. Yu, X. Mineralogical Characteristics and Geological Significance of Spodumene in Lijiagou Pegmatite Deposit in Sichuan. Ph.D. Thesis, Chengdu University of Technology, Chengdu, China, 2021.
9. Qiu, C.Z. Study on the Mineralogical Characteristics of Spodumene in Altai Pegmatites in Xinjiang. Ph.D. Thesis, China University of Geosciences, Beijing, China, 2014.
10. Luo, X.P.; Zhang, Y.B.; Zhou, H.P.; Xie, F.X.; Yang, Z.Z.; Zhang, B.Y. Flotation separation of spodumene and albite with activation of calcium ion hydrolysate components. *Rare Metals* **2022**, *41*, 3919–3931. [CrossRef]
11. Meng, J.P.; Xu, L.H.; Wang, D.H.; Xue, K.; Luo, L.P.; Shi, X.Z. The activation mechanism of metal ions on spodumene flotation from the perspective of in situ ATR-FTIR and ToF-SIMS. *Miner. Eng.* **2022**, *182*, 107567. [CrossRef]

12. Zhang, Y.b.; Zhou, H.p.; Cao, Y.J.; Luo, X.p.; Xie, F.X.; Zhang, B.Y.; Yang, S.Q. Activation mechanism of calcium hydrolysate on the spodumene surface and its effect on the adsorption of collector. *Miner. Eng.* **2021**, *174*, 107221.
13. Li, L.C.; Su, J.; Lu, T.J.; Shen, M.D.; Zhang, J. Luminescence characteristics of green spodumene. In Proceedings of the 2013 China Jewelry Academic Exchange Conference, Beijing, China, 30 October 2013; pp. 94–97. (In Chinese).
14. Sun, C.Y. Differences in the floatiness of different colors of spodumene and crystal chemical analysis. *Nonferrous Metal.* **2000**, *4*, 107–110+103.
15. Rehman, H.U.; Martens, G.; Tsai, Y.L.; Chankhantha, C.; Kidkhunthod, P.; Shen, A.H. An X-ray Absorption Near-Edge Structure (XANES) Study on the Oxidation State of Chromophores in Natural Kunzite Samples from Nuristan, Afghanistan. *Minerals* **2020**, *10*, 5. [CrossRef]
16. Rossman, G.R. The Geochemistry of Gems and Its Relevance to Gemology: Different Traces, Different Prices. *Elements* **2009**, *5*, 159–162. [CrossRef]
17. Ito, A.S.; Isotani, S. Heating effects on the optical absorption spectra of irradiated, natural spodumene. *Radiat. Eff. Defects Solids* **1991**, *116*, 307–314. [CrossRef]
18. Sun, C.Y. Analysis of the difference in planktony of spodumene and neonite in the same ore body. *J. China Univ. Min. Technol.* **2001**, *6*, 5–10.
19. Zhou, H.P.; Xie, F.X.; Zhang, Y.B.; Zhang, B.Y.; Yang, S.Q.; Luo, X.P. Insights into the floatability between spodumene and albite from crystal chemistry standpoint. *Int. J. Min. Sci. Technol.* **2022**, *32*, 1329–1339. [CrossRef]
20. Czaja, M.; Lisiecki, R.; Kadziolka-Gawel, M.; Winiarski, A. Some Complementary Data about the Spectroscopic Properties of Manganese Ions in Spodumene Crystals. *Minerals* **2020**, *10*, 554. [CrossRef]
21. Han, B. Spectroscopic Analysis and Color Optimization of Kunzite in Endeavan. Ph.D. Thesis, China University of Geosciences, Wuhan, China, 2021.
22. Iqbal, M.; Saeed, A.; Zafar, S.I. FTIR spectrophotometry, kinetics and adsorption isotherms modeling, ion exchange, and EDX analysis for understanding the mechanism of Cd^{2+} and Pb^{2+} removal by mango peel waste. *J. Hazard. Mater.* **2008**, *164*, 161–171. [CrossRef]
23. Cui, L.; Guo, Y.; Tang, J.; Yang, Y. Spectroscopy Characteristics and Color-Influencing Factors of Green Iron-Bearing Elbaite. *Crystals* **2023**, *13*, 1461. [CrossRef]
24. Pasetti, L.; Borromeo, L.; Bersani, D.; Andò, S.; Schnellrath, J.; Hennebois, U. Identification of Some Gem Quality Blue to Green Li-Tourmalines. *Minerals* **2024**, *14*, 44. [CrossRef]
25. Wesełucha-Birczyńska, A.; Słowakiewicz, M.; Natkaniec-Nowak, L.; Proniewicz, L.M. Raman microspectroscopy of organic inclusions in spodumenes from Nilaw. *Spectrochim. Acta Part A Mol. Biomol. Spectrosc.* **2011**, *79*, 789–796. [CrossRef] [PubMed]
26. Moon, K.S.; Fuerstenau, D.W. Surface crystal chemistry in selective flotation of spodumene ($LiAl[SiO_3]_2$) from other aluminosilicates. *Int. J. Miner. Process.* **2003**, *72*, 11–24. [CrossRef]
27. Li, Y.L.; Wang, R.P.; Zhu, G.L.; Li, G.S.; Cao, Y.J. A DFT study on the effect of lattice defects on the electronic structures and floatability of spodumene. *Phys. B Condens. Matter* **2024**, *676*, 415657. [CrossRef]
28. Isotani, S.; Watari, K.; Mizukami, A.; Bonventi, W.; Ito, A.S. UV optical absorption spectra analysis of spodumene crystals from Brazil. *Phys. B Condens. Matter* **2007**, *391*, 322–330. [CrossRef]
29. Hu, Y. Effects of Organic Inhibitors in Different Collector Systems on the Flotation Behavior of Spodumene and Gangue Minerals. Ph.D. Thesis, Central South University, Changsha, China, 2023.
30. Jiang, W. Kinetic Simulation of Spodumene Adsorbent Molecules. Ph.D. Thesis, Jiangxi University of Science and Technology, Ganzhou, China, 2015.
31. Yu, F.S.; Yan, P.K.; Jiang, M.; Wang, J.L.; Li, J.; An, F.W. Genetic characteristics of spodumene and potassium feldspar minerals and their floatability. *Met. Mines* **2020**, *06*, 75–80.
32. Yu, F.S.; Sun, Y.F.; Jiang, M.; Xu, G.D.; Wang, J.L.; Li, J. Activation behavior and mechanism of metal cations in spodumene flotation. *Chin. J. Nonferrous Met.* **2021**, *31*, 203–210.
33. Li, S.Y.; Liu, J.; Han, Y.X.; Zhang, S.M. A review of pegmatite-type lithium and beryllium rare metal ores and their flotation separation in China. *Met. Mines* **2022**, *558*, 129–138. [CrossRef]
34. Xie, R.Q.; Zhu, Y.M.; Liu, J.; Li, Y.J. Study on spodumene crystal structure and chemical gene characteristics of (110) surface surface based on density functional theory. *Met. Mines* **2020**, *6*, 68–74.
35. Xu, L.H.; Dong, Q.F.; Liu, J.; Wang, Z.; Wang, J.M. Separation of surface crystal chemistry of diaspore monohydrate and kaolinite by anionic collector flotation. *Miner. J.* **2016**, *36*, 265–270.
36. Zheng, H.T.; Wang, Y.H.; Zhao, Y.H.; Zhu, G.L.; Lu, D.F.; Zheng, X.Y. The influence of NaOH and mechanical stirring on the surface of spodumene and its flotation behavior. *Nonferrous Metall. Eng.* **2019**, *9*, 61–68.
37. Xi, Y.; Dai, S.J.; Zhang, Z.J. Effect of Ca^{2+} on magnesite flotation behavior. *Met. Mines* **2019**, *9*, 98–101.
38. Zhang, G.F.; Tang, P.H.; Zhu, Y.G.; Ding, P.; Feng, Y. Solution chemistry of flotation separation of apatite and quartz in Ca^{2+}-Na_2CO_3 system. *Ind. Miner. Process.* **2011**, *40*, 1–4+19.

39. Zeng, W.W.; Liu, X. Effect of calcium and magnesium ions on flotation of feldspar and epidote and its mechanism. *Min. Metall. Eng.* **2022**, *42*, 59–63.
40. Liu, S.; Xie, L.; Liu, J.; Liu, G.Y.; Zhong, H.; Wang, Y.X.; Zeng, H.B. Probing the interactions of hydroxamic acid and mineral surfaces: Molecular mechanism underlying the selective separation. *Chem. Eng. J.* **2019**, *374*, 123–132. [CrossRef]
41. Pirhashemi, M.; Habibi-Yangjeh, A. Ultrasonic-assisted preparation of plasmonic ZnO/Ag/Ag_2WO_4 nanocomposites with high visible-light photocatalytic performance for degradation of organic pollutants. *J. Colloid Interface Sci.* **2017**, *491*, 216–229. [CrossRef]

minerals

MDPI

Article

Interaction and Inhibition Mechanism of Sulfuric Acid with Fluorapatite (001) Surface and Dolomite (104) Surface: Flotation Experiments and Molecular Dynamics Simulations

Aoao Chen [1], Xuming Wang [2] and Qin Zhang [3,4,5,*]

[1] College of Resources and Environmental Engineering, Guizhou University, Guiyang 550025, China; 18212106728@163.com
[2] College of Mines & Earth Sciences, University of Utah, Salt Lake City, UT 84112, USA; x.wang@utah.edu
[3] Guizhou Academy of Sciences, Guiyang 550001, China
[4] National & Local Joint Laboratory of Engineering for Effective Utilization of Regional Mineral Resources from Karst Areas, Guiyang 550025, China
[5] Guizhou Key Laboratory of Comprehensive Utilization of Non-Metallic Mineral Resources, Guiyang 550025, China
* Correspondence: zq6736@163.com

Abstract: The natural wettability of apatite and dolomite and the effect of sulfuric acid (H_2SO_4) and sodium oleate (NaOl) on the floatability and wettability of both minerals were studied using single-mineral flotation and contact angle measurement. The flotation experiments demonstrated that adding NaOl, apatite, and dolomite had good floatability. After adding H_2SO_4, the floatability of apatite decreased significantly. H_2SO_4 effectively inhibits apatite flotation. Contact angle measurements show that the use of H_2SO_4 induces a significant difference in surface wettability between apatite and dolomite. The moderate addition of H_2SO_4 can increase the contact angle of dolomite. In order to study the selective inhibition mechanism of H_2SO_4 in phosphorite flotation, molecular dynamics simulations (MDSs) were conducted to investigate the interaction between H_2SO_4 and fluorapatite and dolomite at the atomic–molecular level. The results of MDSs reveal that H_2SO_4 interacts with Ca sites on both fluorapatite and defective dolomite surfaces, hindering the interaction of NaOl with Ca sites on both mineral surfaces. SO_4^{2-} ions cannot prevent the interaction of oleate ions with Mg sites on dolomite surface. It is worth mentioning that SO_4^{2-} ions occupy the defective vacancies formed due to the dissolution of CO_3^{2-} on the surface of dolomite and interact with Ca sites. The remaining H_2SO_4 is subsequently adsorbed onto the surface of dolomite. Experimental and simulation results show that, due to the interaction of H_2SO_4 and NaOl, the surface of apatite can still undergo hydration forming a water molecule layer and maintaining a macroscopic hydrophilic property. In contrast, the oleate ions form an adsorption layer on dolomite transitioning it from a hydrophilic to a hydrophobic state. During the phosphate flotation process, the addition of an appropriate amount of sulfuric acid can further diminish the hydration of the dolomite surface, so that the surface of dolomite is more hydrophobic.

Keywords: fluorapatite; dolomite; sulfuric acid; molecular dynamics simulation

1. Introduction

Phosphate ore is a crucial raw material for the manufacturing of phosphoric acid. It is primarily utilized in the production of fertilizers and various other phosphorus-containing products, including yellow phosphorus, food additives, and phosphorus-based chemical compounds [1]. With the rapid growth in the world's population, the demand for phosphate ore in both agricultural production and industry has been steadily increasing. It is reported that approximately 82% of the world's phosphate ore is used to produce phosphate fertilizers, while the remaining portion is used in industries such as metallurgy, materials, pharmaceuticals, and light industry [2,3]. China is rich in phosphate ore resources, with phosphate ore reserves ranking second in the world [4].

However, it is important to note that phosphate rock resources are characterized by their abundance rather than their richness in China. Most of these resources consist of apatite [5], a mineral often associated with gangue minerals like dolomite. Apatite exhibits a complex structure, fine disseminated particle size, and poses challenges in terms of dissociation [6,7]. The separation of apatite from dolomite is notably challenging, primarily due to their similar surface physical and chemical properties [8].

Due to the low grade of apatite, it cannot be directly used to produce phosphate industrial products, and it requires enrichment through a mineral processing process to meet production requirements. At present, many separation methods are being applied to phosphate ore dressing, but froth flotation remains the most widely used process [9].

In flotation systems, the adsorption of agents onto the mineral surface plays a crucial role in mineral separation. Fatty acid reagents are commonly used as collectors in phosphate ore flotation. However, a challenge arises because both apatite and gangue minerals (such as dolomite, calcite) are calcium minerals, resulting in a similar exposure of their active sites in solution. This similarity leads to poor selectivity of these fatty-acid-type collectors in the flotation of calcium-bearing minerals [10,11]. Directly adding fatty acid collectors cannot effectively separate apatite from dolomite. Therefore, inhibitors must be introduced to facilitate the separation of apatite from other calcium-containing gangue minerals through flotation. In order to achieve efficient separation between apatite and carbonate minerals, extensive research has been conducted on inhibitors.

Liu et al. [12] used 2-phosphobutane-1,2,4-tricarboxylic acid (PBTCA) as an inhibitor to separate apatite and calcite. The flotation results showed that PBTCA reduced the adsorption of NaOl on the surface of calcite, achieving effective separation of apatite and calcite. Wang et al. [13] utilized carboxymethyl chitosan as a selective inhibitor to separate apatite and calcite. Their research found that the -COO^- of the inhibitor was adsorbed on the surface of calcite through chemical chelation at the Ca site, while carboxymethyl chitosan was adsorbed on the surface of apatite through hydrogen bonding. Both apatite and carboxymethyl chitosan were negatively charged, resulting in strong electrostatic repulsion and steric hindrance between carboxymethyl chitosan and apatite. This made it more difficult for carboxymethyl chitosan to adsorb on the surface of apatite. Zhong et al. [14] used sodium alginate as an inhibitor to separate apatite and dolomite. The flotation results showed that, within the pH range of 8–11, sodium alginate could completely inhibit dolomite, while having a minimal effect on apatite flotation.

Depending on the type of accompanying gangue minerals, different flotation processes are used to separate phosphate rock from these gangue minerals. For phosphate rock with gangue minerals mainly consisting of dolomite, reverse flotation is the most effective and widely used method of separating dolomite from phosphate rock [15]. In practical production, H_2SO_4 is commonly used as an inhibitor, and then fatty acid collectors are used in weakly acidic environments to separate carbonate gangue minerals such as dolomite from apatite [6]. Currently, there is a greater focus on researching the interaction between H_2SO_4 and apatite [16,17], but the interaction mechanism between H_2SO_4 and dolomite remains unclear. In recent years, computer simulation technology has advanced rapidly, and using molecular simulation to characterize the adsorption process and interface interactions of reagents on mineral particle surfaces has proven to be an effective research method [18]. For instance, Xie et al. [19] used DFT (density functional theory) to study the effects of carbon chain length, carbon chain configuration, and the number of C=C double bonds on the surface of fluorapatite (FAP). Their results showed that the fatty acid collector formed a stable adsorption configuration on the FAP surface, with the O of the fatty acid collector undergoing chemical adsorption with Ca1 on the FAP (001) surface. The H of the fatty acid collector underwent hydrogen bond adsorption with O on the FAP (001) surface. Molecular simulation has the capability to depict phenomena that are not observable in experiments at the molecular atomic scale, thus providing a better explanation for certain macroscopic experimental results. In addition, the results of molecular simulation can offer valuable guidance for our experimental research.

In this paper, the interactions between H_2SO_4 and the surfaces of apatite and dolomite were elucidated at the molecular–atomic level through molecular dynamics simulations using fluorapatite (001) and dolomite (104) as the objects of this study, which revealed the mechanism of selective inhibition of H_2SO_4 and elucidated the interactions between H_2SO_4, NaOl, and apatite and dolomite.

2. Materials and Methods

2.1. Mineral Samples

The experimental ore sample is a medium–low-grade calcium–magnesium phosphate ore taken from a phosphate mine in Guizhou, China. As can be seen from Figure 1, the valuable mineral in the original ore is fluorapatite ($Ca_5F(PO4)_3$), while dolomite ($CaMg(CO_3)_2$) is the associated vein mineral.

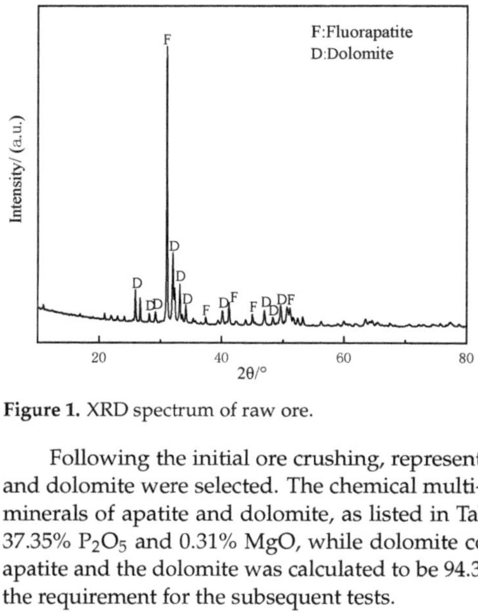

Figure 1. XRD spectrum of raw ore.

Following the initial ore crushing, representative single-mineral samples of apatite and dolomite were selected. The chemical multi-element analysis results of these single minerals of apatite and dolomite, as listed in Table 1, indicate that the apatite contains 37.35% P_2O_5 and 0.31% MgO, while dolomite contains 20.83% MgO. The purity of the apatite and the dolomite was calculated to be 94.31% and 95.81%, respectively, which met the requirement for the subsequent tests.

Table 1. Single-mineral chemical multi-element analysis/%.

Element	P_2O_5	CaO	MgO	SiO_2	Fe_2O_3	Al_2O_3	K_2O	Na_2O	Others
Apatite	37.35	46.31	0.31	0.77	0.09	1.17	0.01	0.06	13.93
Dolomite	0.08	26.54	20.83	2.81	0.02	0.75	0.00	0.09	48.88

2.2. Contact Angle Measurement

Large pieces of collophane and dolomite samples were selected and initially polished using sandpaper. Subsequently, the polished specimens were immersed in distilled water and subjected to ultrasonic cleaning to ensure that the polished surfaces were free from contaminants.

For contact angle measurements, the sessile drop method was employed. The test specimen was immersed in a reagent solution of the desired concentration for 10 min and then dried. The sessile drop contact angle was then measured. Three measurements were carried out for each reagent solution condition, and the average value was reported.

2.3. Single-Mineral Flotation

The single-mineral flotation experiments involving apatite and dolomite were conducted using an XFGC II inflatable hanging cell flotation machine manufactured by Jilin Prospecting Machinery Factory, China. The mixing speed was set at 1992 RPM. In each test, 2 g of individual minerals and 40 mL of deionized water were placed into a flotation cell. Prior to adding the reagents, the pulp was stirred for 30 s to ensure the proper dispersion of mineral particles in the deionized water. Subsequently, the inhibitor was added, followed by a 30 s conditioning period, after which the collector was introduced before initiating the flotation process. The floated product was manually skimmed off from the top of the cell. The resulting flotation process lasted for 5 min. The products within the flotation cell and the floated products were then filtered, dried, and weighed to calculate the flotation recovery. Each flotation condition experiment was repeated three times under identical experimental conditions, and the average value was determined.

2.4. Molecular Dynamics Simulation

2.4.1. Simulation Method

The Forcite module in MS software was used for MDS calculations. For all simulation calculations, the COMPASSII force field, NVT ensemble, and Velocity Scale thermostat methods were used. Atom-based and Ewald methods were used to calculate van der Waals and electrostatic interactions, with an accuracy of Fine. During dynamic simulation calculations on the system, the lower layer of fluorapatite and dolomite were immobilized, and a simulation time step of 1.0 fs was used to calculate the motion equations. The total simulation time for each simulation process was 500 ps, and after the simulation was completed, the final 100 ps trajectory file was selected for the calculation and analysis of the relative concentration distribution.

2.4.2. Simulation Model

The crystal models of both the minerals were sourced from the AMCSD database. The original lattice parameters for fluorapatite were a = b = 9.397 Å and c = 6.878 Å, while the original lattice parameters for dolomite were a = b = 4.815 Å and c = 16.119 Å. After optimization, the lattice parameters for fluorapatite became a = b = 9.444 Å and c = 6.887 Å, and, for dolomite, they were a = b = 4.823 Å and c = 16.046 Å. These optimized lattice parameters closely matched the original lattice parameters. To prepare the system, the fluorapatite and dolomite mineral crystals were cleaved along their most common cleavage planes, expanding the supercell. The Amorphous Cell module was used to construct the solution model of the pharmaceutical solution. In addition, the Build Layers tool was used to construct the pharmaceutical solution model and incorporated fluorapatite or dolomite to form the fluorapatite–/dolomite–pharmaceutical solution system. The number of water molecules in the solution model was 2000; the number of H_2SO_4 molecules was 100; and the number of NaOl molecules was 20. The optimized configurations for sulfate and oleic acid ions are shown in Figure 2.

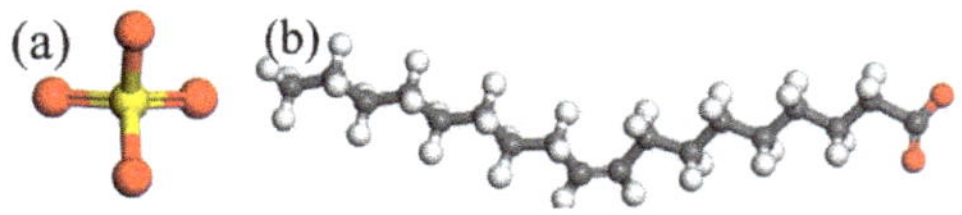

Figure 2. The optimized configurations of SO_4^{2-} and oleate ion. (**a**) SO_4^{2-}; (**b**) oleate ion.

To prevent any effects from periodic boundary conditions, a vacuum section with a thickness of 20 Å was added at the top of the system. Subsequently, the constructed system was geometrically optimized and dynamically simulated using the Forcite module to obtain the lowest-energy-state configuration. The constructed model is shown in Figure 3.

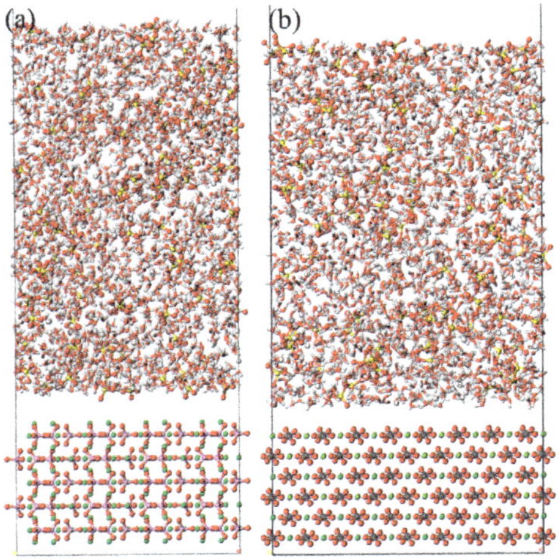

Figure 3. The constructed mineral–H_2SO_4 solution system model. (**a**) Fluorapatite–H_2SO_4 solution system; (**b**) dolomite–H_2SO_4 solution system.

3. Results and Discussion

3.1. Study on Flotation Behavior of Apatite and Dolomite

The impact of various dosages of NaOl on the flotation ratio of apatite and dolomite is depicted in Figure 4. From the effect of NaOl dosage on the floatation ratio of apatite and dolomite shown in Figure 4a, it follows that the floatation ratio of apatite and dolomite increases with an increase in NaOl dosage, which indicates that NaOl has good collection performance for apatite and dolomite. The agent has no selectivity for apatite and dolomite. When the dosage of NaOl was 150 mg/L, the floatation ratio of apatite and dolomite exceed 95%. Therefore, the appropriate dosage of NaOl was chosen as 150 mg/L. When the dosage of NaOl was below 150 mg/L, the floatation ratio of apatite was higher than that of dolomite. This could be due to the superior floatation properties of apatite compared with dolomite. When the dosage of NaOl exceeded 150 mg/L, the floatation ratio of apatite showed a downward trend.

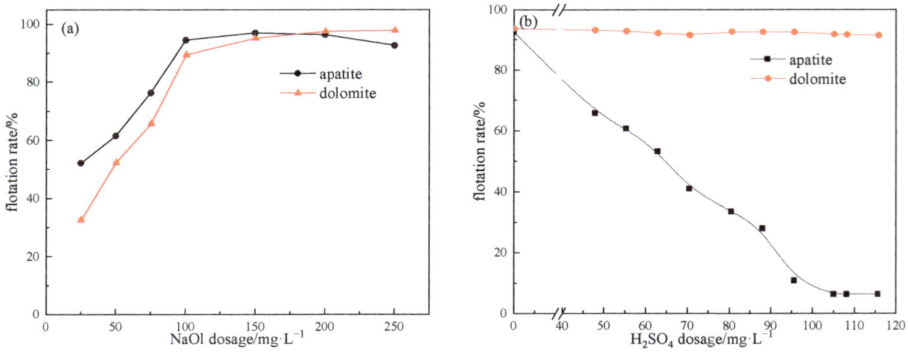

Figure 4. Effect of agent dosage on the floatation ratio of apatite and dolomite. (**a**) NaOl dosage. (**b**) H_2SO_4 dosage; the dosage of NaOl was 150 mg/L.

The floatation ratio of apatite and dolomite under different H_2SO_4 dosages with 150 mg/L NaOl is shown in Figure 4b. It can be seen that, without the addition of H_2SO_4, the floatation ratios of apatite and dolomite were very close, with both having higher floatation ratios of over 90%. The addition of H_2SO_4 resulted in a decrease in the floatation ratio of apatite, while the floatation ratio of the dolomite was essentially unchanged. As the dosage of H_2SO_4 increased, there was a continued decrease in the floatation ratio of apatite. In the presence of 110 mg/L H_2SO_4, the floatation ratios of apatite and dolomite were 6.38% and 91.73%, which represents a decrease of 86.16% and 1.93% compared that with the absence of H_2SO_4. The decreasing trend in the apatite flotation ratio slows down when the H_2SO_4 dosage exceeds 100 mg/L. The apatite flotation ratio remained essentially unchanged as the H_2SO_4 dosage continued to increase. The effect of H_2SO_4 on the floating ratio of dolomite was slight, and the floating ratio of dolomite remained above 90%.

3.2. Effect of Reagent on Wettability of Mineral Surface

Figure 5 shows the effect of H_2SO_4 and NaOl on the wettability of apatite and dolomite. Without adding any reagents, the contact angle of apatite was 56° and that of dolomite was 61°. The natural contact angles of apatite and dolomite show that they have good hydrophilicity. The natural contact angle of apatite, tested before, is about 10°. The contact angle of apatite tested in this study is greater than the theoretical value for apatite or dolomite. The apatite and dolomite used in the experiment were derived from sedimentary phosphate rock; in the complex sedimentary environment, some minerals, such as fluorapatite and dolomite, frequently undergo intricate ion exchange, organic matter or other impurity deposition, and infection. These conditions may be responsible for the increased contact angle of both minerals [20].

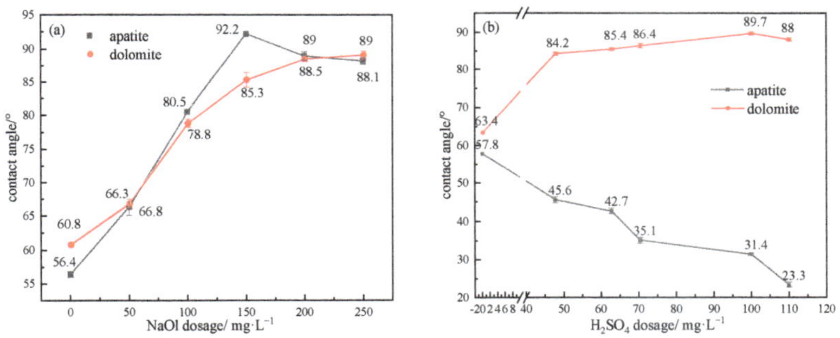

Figure 5. Effect of NaOl and H_2SO_4 dosage on the wettability of apatite and dolomite, (**a**) Effect of NaOl dosage on the wettability of apatite and dolomite. (**b**) Effect of H_2SO_4 dosage on the wettability of apatite and dolomite; the dosage of NaOl was 150 mg/L.

As shown in Figure 5a. The contact angle of apatite and dolomite increased with an increasing dosage of NaOl when only NaOl was added. This shows that the surface of apatite and dolomite changed from hydrophilic to hydrophobic under the action of NaOl. The hydration of the surface of apatite and dolomite can be improved by NaOl. When the dosage of the collector was 150 mg/L, the contact angle of the apatite was 92°, and the contact angle of the dolomite was 85°. Compared with the situation without the addition of the reagent, the contact angle of apatite and dolomite increased by 36° and 24°. When the dosage of NaOl was more than 150 mg/L, the contact angle of the apatite decreased. The main reason is that NaOl forms a large cylindrical micelle adsorption layer with a bimolecular layer on the surface of apatite when the amount of NaOl exceeds a certain concentration. The formation of multilayered large cylindrical micelles led to a decrease in the hydrophobicity of the collophane surface [21].

Figure 5b shows that, at a NaOl dosage of 150 mg/L, the contact angle of apatite decreased as the H_2SO_4 dosage increased, while the contact angle of dolomite increased slightly. At a dosage of 110 mg/L of H_2SO_4, the contact angles of collophane and dolomite were 23.3° and 88°, respectively. The contact angle difference between both minerals was 64.7°. The adsorption of NaOl on the surface of apatite was hindered by H_2SO_4. As a consequence, the surface of apatite remained hydrophilic. Following the adsorption of

H_2SO_4 on the surface of dolomite, NaOl could still adsorb on the surface of dolomite to render it hydrophobic. The interaction of H_2SO_4 and NaOl can selectively regulate the hydration of apatite and dolomite. It is worth noting that the contact angle of dolomite was found to increase slowly as the H_2SO_4 dosage was increased during the test. The contact angle of dolomite was the largest at a H_2SO_4 dosage of 100 mg/L. When the H_2SO_4 dosage was further increased, the contact angle of dolomite decreased, but it was still higher than the previous contact angle. The contact angle of dolomite can be increased by adding H_2SO_4, and the hydrophobicity of dolomite can be improved. Zou et al. [17] also observed the same phenomenon in the contact angle test in their study, in which an increase in H_2SO_4 dosage increased the contact angle of dolomite, but they did not provide a reasonable explanation. The test results are consistent with the contact angle test results in Zou's research.

3.3. Molecular Dynamics Simulation of Fluorapatite–/Dolomite–Agent Solution System

Figure 6 shows the molecular dynamics configuration of the fluorapatite–/dolomite–H_2SO_4 system. The CO_3^{2-} on the surface of dolomite is unstable in acid solution. In H_2SO_4 solution, CO_3^{2-} on the surface of dolomite will dissolve into CO_2 and H_2O. There will be defects on the surface of dolomite [22]. The adsorption of SO_4^{2-} on the perfect dolomite surface and the defective dolomite surface was compared. According to the equilibrium configuration and bond population calculated using the system's molecular dynamics, theO atoms in SO_4^{2-} react with the Ca atoms on the fluorapatite surface to form Ca-O bonds with partial bond lengths of 1.817 and 1.868 Å. As shown in the black circle in Figure 6a. On the perfect dolomite surface, SO_4^{2-} did not bond with Ca/Mg sites, and SO_4^{2-} did not adsorb on the dolomite surface. On defective dolomite surfaces, SO_4^{2-} interacted with Ca sites with partial Ca-O bond lengths of 1.856 and 2.109 Å. As shown in the black circle in Figure 6c. The simulation results of the MDS indicate that SO_4^{2-} interacted with the Mg sites. Nevertheless, $MgSO_4$ displayed a high solubility [17,22]. In the real solution, the Mg present on the surface of dolomite reacted with SO_4^{2-} to produce $MgSO_4$, which then dissolved.

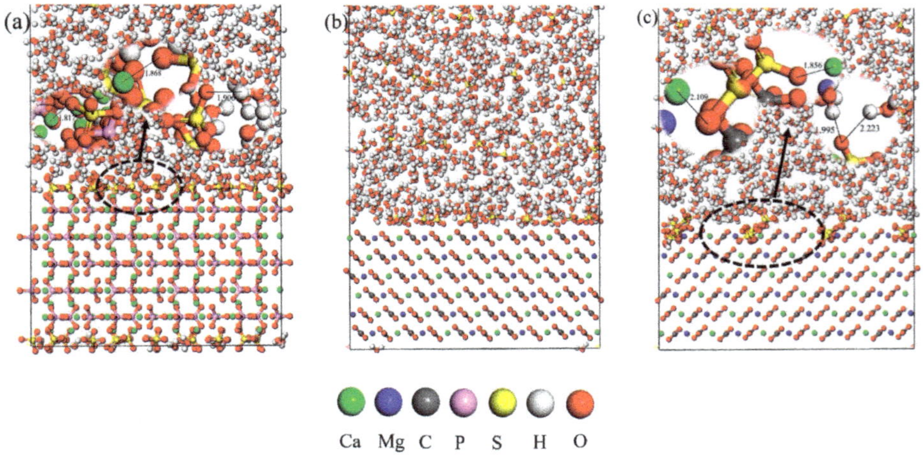

Figure 6. Molecular dynamics calculation of equilibrium configuration of mineral–H_2SO_4 solution system. (**a**) Fluorapatite; (**b**) perfect dolomite; (**c**) defective dolomite.

The SO_4^{2-} interacted with fluorapatite and defective dolomite but not with the perfect dolomite surface. According to the adsorption configuration of SO_4^{2-} on the surface of defective dolomite, SO_4^{2-} mainly occupied the defect position produced by the dissolution of CO_3^{2-} on the surface of dolomite and acted on the Ca site, the remaining SO_4^{2-} was adsorbed on the dolomite surface. The SO_4^{2-} adsorption amount on the dolomite surface was lower than that on the fluorapatite surface. The interaction between H_2SO_4 and dolomite was further demonstrated by molecular dynamics simulations. The SO_4^{2-} formed hydrogen bonds with water molecules in the solution, with some bond lengths of 1.906, 1.995, and 2.223 Å. Zou et al. [17] used TOF-SIMS to measure the surface of H_2SO_4 treated dolomite and also found the presence of SO_4^{-} on the surface, indicating that SO_4^{2-} can adsorb on the surface of dolomite. TOF-SIMS is a highly sensitive technique to determine the 2D/3D distributions of chemical fragments on a solid surface. The molecular dynamics simulation results are consistent with their test results.

The interfacial adsorption energies of fluorapatite and dolomite with H_2SO_4 were calculated, and the results are listed in Table 2. The adsorption energy at the fluorapatite–H_2SO_4 interface was $-117{,}830.60$ kcal·mol^{-1}; at the perfect dolomite–H_2SO_4 interface, it was 1.90 kcal·mol^{-1}, and at the defective dolomite–H_2SO_4 interface it was $-100{,}334.92$ kcal·mol^{-1}. The adsorption energies at the fluorapatite–/defective dolomite–H_2SO_4 interface were all negative, and the adsorption energy at the perfect dolomite–H_2SO_4 interface was positive. It indicates that H_2SO_4 can spontaneously adsorb on the surfaces of fluorapatite and defective dolomite. Cao et al. [23] studied the interactions of individual sulfate molecules with dolomite by using the density flood theory and also found that sulfate adsorption at the surface of perfect dolomite is not an energy-favorable process. The results of the present study are in agreement with the results obtained by them. The SO_4^{2-} does not interact with perfect dolomite. In actual flotation, perfect crystal surfaces are rare after the ore has been crushed, milled, and subjected to the action of H_2SO_4 in the flotation process. Therefore, defective dolomite surfaces were chosen for all subsequent simulations.

Table 2. Fluorapatite–/dolomite–H_2SO_4 interface adsorption energy.

System	E_{ads}/kcal·mol^{-1}
Fluorapatite–H_2SO_4	$-117{,}830.60$
Perfect dolomite–H_2SO_4	1.90
Defective dolomite–H_2SO_4	$-100{,}334.92$

Figure 7 shows the configuration of the fluorapatite–/dolomite–H_2SO_4 and NaOl solution system after the molecular dynamics calculations were completed. According to the kinetic calculation equilibrium configuration and bond population analysis, it can be seen that the Ca site on the surface of fluorapatite was not bonded with the oleate ions. The main components on the surface of fluorapatite were SO_4^{2-} and H_2O. A large number of oleate ions were adsorbed on the surface of the dolomite, and the Mg-O bonds were formed by the interaction of oleate ions with Mg atoms on the surface of the dolomite. Some of the Mg-O bonds were 1.610 and 1.680 Å in length, and as shown in the black circle in Figure 7b. The surface of dolomite was mainly composed of SO_4^{2-} and oleate ions.

As shown in Table 3, the interfacial adsorption energies of fluorapatite–/dolomite–oleate ions were 4312.55 and $-30{,}891.66$ kcal·mol^{-1}, respectively. The interfacial adsorption energy of fluorapatite–oleate ions also provided further evidence that oleate ions did not interact with the fluorapatite surface. The interaction of oleate ions with the Ca sites on the fluorapatite surface was prevented by the interaction of H_2SO_4 with the Ca sites. The interaction of oleate with the Mg sites on the dolomite surface was not affected by H_2SO_4.

The trajectory files obtained after the molecular dynamics simulation can be calculated and analyzed for the relative concentrations and root mean square displacements (MSD) of water molecules and agent molecules on the mineral surface, which can be used to characterize the adsorption structure, kinematic properties, and adsorption tightness of agent molecules on the mineral surface [24–26].

Figure 8 shows the relative concentration profiles of SO_4^{2-} in the H_2SO_4 solution on the surface of the fluorapatite and dolomite along the z-axis, which were calculated from the trajectory files at the end of the simulation. The crystal thicknesses of fluorapatite and dolomite were 1.87 and 1.71 nm. The analysis of the relative concentration distribution showed that the first peak of SO_4^{2-} on the surface of fluorapatite appeared at 2.05 nm along the z-axis on the surface of fluorapatite, which indicates that sulfate adsorbed on the surface of fluorapatite. On the dolomite surface, the first peak of the SO_4^{2-} concentration distribution curve appeared at 1.64 nm along the z-axis. The first peak of SO_4^{2-} concentration appeared in the dolomite crystal, and there were two obvious peaks for SO_4^{2-}. It is further demonstrated that SO_4^{2-} first occupied the defective vacancies created by the dissolution of CO_3^{2-} and bound to the exposed Ca sites. The remaining SO_4^{2-} was adsorbed on the dolomite surface. The relative concentration distribution curve of water molecules can be used to characterize the intensity of hydration and the thickness of the hydration film formed on the mineral surface [24]. There was a peak in the relative concentration distribution curve of water molecules after the action of H_2SO_4. This demonstrates that hydration was present on the surfaces of fluorapatite and dolomite and that both minerals were capable of forming hydrated films on their surfaces.

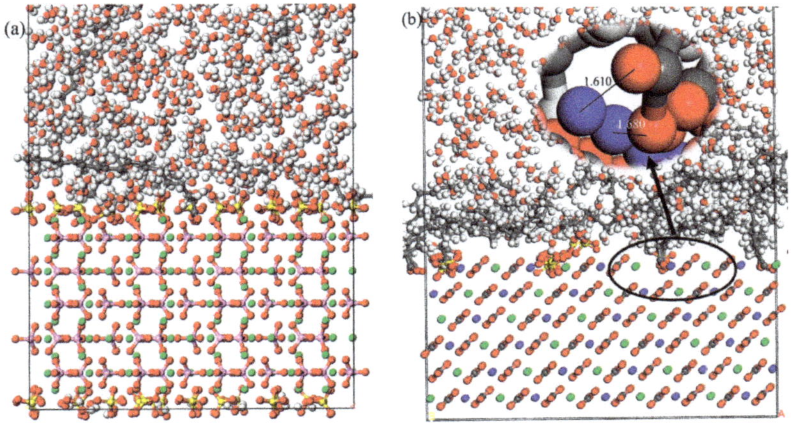

Figure 7. Equilibrium configuration of fluorapatite–/dolomite–H_2SO_4 and NaOl solution system. (**a**) Fluorapatite; (**b**) dolomite.

Table 3. Adsorption energy of fluorapatite–/dolomite–oleate ion interface.

System	E_{ads}/kcal·mol^{-1}
Fluorapatite–oleate ion	4312.55
Defect dolomite–oleate ion	−30,891.66

Figure 9 shows the relative concentration distribution curves of SO_4^{2-}, oleate ions, and water molecules in H_2SO_4 and NaOl solutions on the surface of both minerals along the z-axis direction. The peak of the relative concentration distribution curve of oleate ions on the dolomite surface was very high. Combined with the analysis of the adsorption configuration, it can be seen that the hydrophobic end of the hydrophobicity of the oleate ions faced upward after interacting with the Mg sites on the dolomite surface. In the presence of high concentrations of oleate ions in the solution, the oleate ions interacted with the Mg sites on the dolostone surface to form an adsorbent layer, rendering the dolostone hydrophobic.

The relative concentrations of water molecules on the surface of the fluorapatite peak are shown in Figure 9. The fact that the water molecules were denser on the surface of fluorapatite is evident from the adsorption configuration of both systems. The results showed that the surface of fluorapatite was still hydrated after interaction with H_2SO_4 and NaOl. The relative concentration distribution curve of water molecules on the surface of dolomite had no peak value, and there was no hydration on the surface of dolomite after the adsorption of oleate ions. The relative concentration distribution curves explain the decrease in the floatation ratio and contact angle of apatite after the addition of H_2SO_4 in the single-mineral flotation and the contact angle measurement. The simulation results were in agreement with the results of single-mineral flotation and contact angle measurement. In the reverse flotation of phosphate rock, the interaction between oleate ions and the surface of apatite was hindered by H_2SO_4 to keep it hydrophilic, and the apatite particles in the pulp were hydrophilic and sank. The interaction between oleate ions and Mg sites on the surface of dolomite was not affected by H_2SO_4. A large amount of oleate ion was adsorbed on the surface of dolomite, which made the surface of dolomite have good hydrophobicity. The dolomite particles adhered to bubbles and floated up; the flotation models of apatite and dolomite are presented in Figure 10.

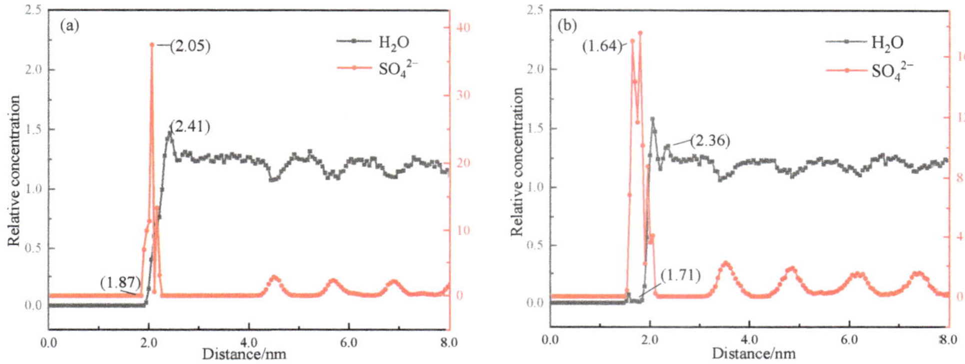

Figure 8. Relative concentration distribution curves of sulfate in fluorapatite–/dolomite–sulfuric acid solution along the z-axis on the mineral surface. (**a**) Fluorapatite; (**b**) dolomite.

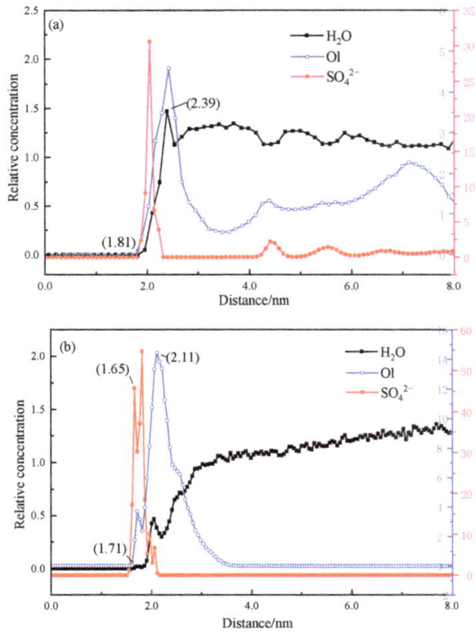

Figure 9. Relative concentrations of SO_4^{2-}, oleate ions, and water molecules along the z-axis at the mineral surface in fluorapatite–/dolomite–sulfuric acid, NaOl solutions. (**a**) Fluorapatite; (**b**) dolomite.

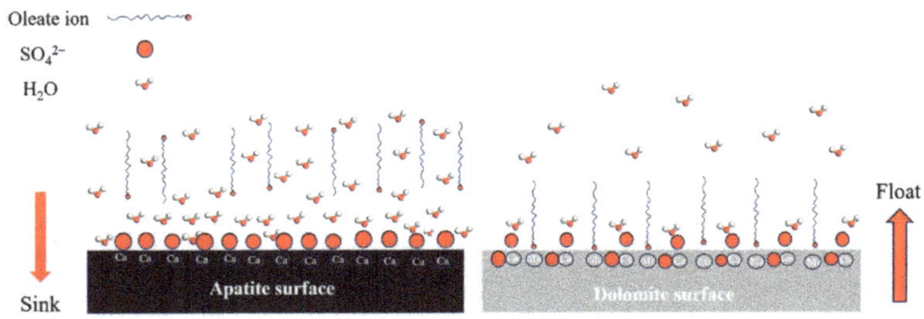

Figure 10. Schematic diagram of the interaction model of H_2SO_4 and NaOl on the surface of apatite and dolomite.

Figure 11 shows the relative concentration distribution curves of water molecules on the dolomite surface along the z-axis direction for different amounts of H_2SO_4. The relative concentration of water molecules on the dolomite surface was larger when the sulfate was less, and the relative concentration of water molecules on the surface decreased on continuing to increase the amount of H_2SO_4. Combining the equilibrium adsorption configuration analysis, when the solution is free of H_2SO_4, all the exposed Ca/Mg sites on the dolomite surface will interact with oleate ions, forming a tight adsorption layer on the surface. In the presence of H_2SO_4 in the solution, the SO_4^{2-} interacts with the Ca sites on the surface of dolomite. After the sulfate ions are adsorbed on the surface of dolomite, the active sites on the surface of dolomite decrease, and the adsorption layer formed by the action of oleate ions becomes looser. Some water molecules can break through the adsorption layer of oleate ions and adsorb on the surface of dolomite. Water molecules will interact with SO_4^{2-} to adsorb onto the surface of the dolomite and water molecules will also interact with SO_4^{2-} adsorbed onto the surface to form hydrogen bonds if the amount of H_2SO_4 in the solution is low. When there is more H_2SO_4 in the solution, the defect space on the dolomite surface is completely occupied, SO_4^{2-} is adsorbed on the dolomite surface, and the SO_4^{2-} adsorbed on the dolomite surface hinders the interaction between water molecules and Ca sites on the dolomite surface; the water molecules can only interact with surface-adsorbed SO_4^{2-} radicals. This may be the reason why the contact angle of dolomite increased with increasing H_2SO_4 content and decreased again with increasing H_2SO_4 content.

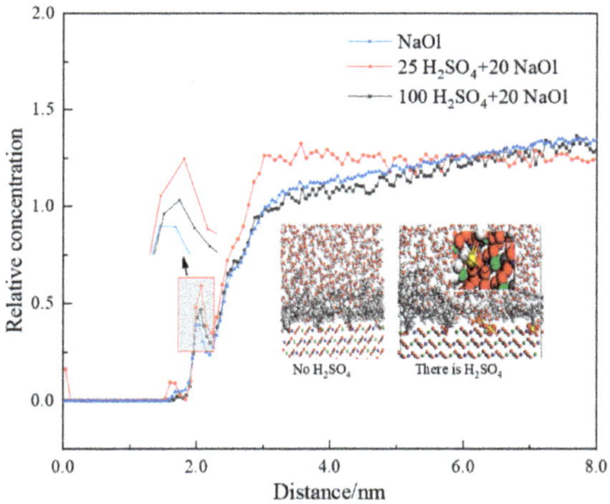

Figure 11. Relative concentration distribution curves of water molecules on the surface of dolomite along the z-axis for different amounts of H_2SO_4.

4. Conclusions

NaOl enhances the floatability and hydrophobicity of apatite and dolomite, but it does not show selectivity between them. H_2SO_4 effectively inhibits apatite, leading to significant differences in the floatability and wettability of both minerals. Following the sulfuric acid treatment, the contact angle of dolomite slightly increases, and H_2SO_4 can interact with dolomite to slightly enhance its hydrophobicity.

The SO_4^{2-} ions can adsorb on the surface of fluorapatite and defective dolomite simultaneously. Initially, SO_4^{2-} ions occupy the defect position after CO_3^{2-} dissolution on the surface of dolomite and then interact with Ca sites. The interaction of SO_4^{2-} ions with the fluorapatite and dolomite surfaces hinders the interaction of the oleate ions with the Ca sites on the surface but does not hinder the interaction of oleate ions with the Mg site on the surface of the dolomite. The simulation results revealed the reasons for the significant differences in the floatability and wettability of the two minerals after the addition of H_2SO_4 and NaOl. The adsorption of H_2SO_4 on the surface of apatite hindered the adsorption of NaOl, maintaining hydration on the surface to form a water molecular layer and keeping it hydrophilic. After the action of H_2SO_4, NaOl can still interact with dolomite. The formation of an oleate ion adsorption layer on the surface of dolomite hindered the formation of the water molecular layer and made it hydrophobic. The concentration of water molecules on the surface of dolomite can be slightly increased after the interaction between SO_4^{2-} ions and the surface of the dolomite, which can well explain the reason for the increase in the contact angle of dolomite after adding H_2SO_4.

Author Contributions: Conceptualization, A.C. and Q.Z.; methodology, A.C.; software, A.C.; validation, A.C., Q.Z. and X.W.; formal analysis, A.C.; investigation, A.C.; resources, Q.Z.; data curation, A.C.; writing—original draft preparation, A.C.; writing—review and editing, A.C., Q.Z. and X.W.; visualization, A.C.; supervision, Q.Z.; project administration, Q.Z.; funding acquisition, Q.Z. All authors have read and agreed to the published version of the manuscript.

Funding: This research was funded by [the National Key R & D Program] grant number [2018YFE0110300].

Data Availability Statement: The data presented in this study are available on request from the corresponding author. The data are not publicly available due to restrictions privacy.

Conflicts of Interest: The authors declare no conflict of interest.

References

1. Liu, X.; Luo, H.; Cheng, R.; Li, C.; Zhang, J. Effect of citric acid and flotation performance of combined depressant on collophanite ore. *Miner. Eng.* **2017**, *109*, 162–168. [CrossRef]
2. Cordell, D.; Drangert, J.O.; White, S. The story of phosphorus: Global food security and food for thought. *Glob. Environ. Chang.* **2009**, *19*, 292–305. [CrossRef]
3. Ruan, Y.; He, D.; Chi, R. Review on beneficiation techniques and reagents used for phosphate ores. *Minerals* **2019**, *9*, 253. [CrossRef]
4. Summaries, M.C. US Geological Survey. Available online: https://pubs.usgs.gov/periodicals/mcs2022/mcs2022.pdf (accessed on 19 May 2022).
5. Xu, W.; Liang, Q.; Tian, Y.; Mei, G. Reverse anionic flotation of dolomitic collophanite using a mixed fatty acid collector: Adsorption behavior and mechanism. *Physicochem. Probl. Miner. Process.* **2022**, *58*, 151519. [CrossRef]
6. Xu, W.; Shi, B.; Tian, Y.; Mei, G. Process mineralogy characteristics and flotation application of a refractory collophanite from Guizhou, China. *Minerals* **2021**, *11*, 1249. [CrossRef]
7. Yang, H.; Xiao, J.; Xia, Y.; Xie, Z.; Tan, Q.; Xu, J.; Guo, H.; Wu, S. Origin of the Ediacaran Weng'an and Kaiyang phosphorite deposits in the Nanhua basin, SW China. *J. Asian Earth Sci.* **2019**, *182*, 103931. [CrossRef]
8. Zhang, H.; Zhou, F.; Liu, M.; Jin, Y.; Yu, H. Employing sulfur—Phosphorus mixed acid as a depressant: A novel investigation in flotation of collophanite. *Energy Sources Part A Recovery Util. Environ. Eff.* **2020**, 1–14. [CrossRef]
9. Deng, J.; Zhang, K.; He, D.; Zhao, H.; Hakkou, R.; Benzaazoua, M. Occurrence of Sesquioxide in a Mid-Low Grade Collophane-Sedimentary Apatite Ore from Guizhou, China. *Minerals* **2020**, *10*, 1038. [CrossRef]
10. Yang, B.; Zhu, Z.; Sun, H.; Yin, W.; Yao, J. Improving flotation separation of apatite from dolomite using PAMS as a novel eco-friendly depressant. *Miner. Eng.* **2020**, *156*, 106492. [CrossRef]
11. Filippova, I.V.; Filippov, L.O.; Lafhaj, Z.; Barres, O.; Fornasiero, D. Effect of calcium minerals reactivity on fatty acids adsorption and flotation. *Colloids Surf. A Physicochem. Eng. Asp.* **2018**, *545*, 157–166. [CrossRef]

12. Liu, C.; Zhang, W.; Li, H. Selective flotation of apatite from calcite using 2-phosphonobutane-1, 2, 4-tricarboxylic acid as depressant. *Miner. Eng.* **2019**, *136*, 62–65. [CrossRef]
13. Wang, T.; Feng, B.; Guo, Y.; Zhang, W.; Rao, Y.; Zhong, C.; Zhang, L.; Cheng, C.; Wang, H.; Luo, X. The flotation separation behavior of apatite from calcite using carboxymethyl chitosan as depressant-ScienceDirect. *Miner. Eng.* **2020**, *159*, 106635. [CrossRef]
14. Zhong, C.; Feng, B.; Zhang, W.; Zhang, L.; Guo, Y.; Wang, T.; Wang, H. The role of sodium alginate in the flotation separation of apatite and dolomite. *Powder Technol.* **2020**, *373*, 620–626. [CrossRef]
15. Abouzeid, A.-Z.M. Physical and thermal treatment of phosphate ores—An overview. *Int. J. Miner. Process.* **2008**, *85*, 59–84. [CrossRef]
16. Liu, X.; Li, C.; Luo, H.; Cheng, R.; Liu, F. Selective reverse flotation of apatite from dolomite in collophanite ore using saponified gutter oil fatty acid as a collector. *Int. J. Miner. Process.* **2017**, *165*, 20–27. [CrossRef]
17. Zou, H.; Cao, Q.; Liu, D.; Yu, X.; Lai, H. Surface features of fluorapatite and dolomite in the reverse flotation process using sulfuric acid as a depressor. *Minerals* **2019**, *9*, 33. [CrossRef]
18. Liu, L.; Qiao, E.; Shen, L.; Min, F.; Xue, C. Effect of hydration layer on the adsorption of dodecane collector on low-rank coal: A molecular dynamics simulation study. *Processes* **2020**, *8*, 1207. [CrossRef]
19. Xie, J.; Li, X.; Mao, S.; Li, L.; Ke, B.; Zhang, Q. Effects of structure of fatty acid collectors on the adsorption of fluorapatite (0 0 1) surface: A first-principles calculations. *Appl. Surf. Sci.* **2018**, *444*, 699–709. [CrossRef]
20. Kholodov, V.N.; Nedumov, R.I. The role of black shales in the formation of phosphate and manganese ores. *Lithol. Miner. Resour.* **2011**, *46*, 321–352. [CrossRef]
21. Ye, J.; Zhang, Q.; Li, X.; Wang, X.; Ke, B.; Li, X.; Shen, Z. Effect of the morphology of adsorbed oleate on the wettability of a collophane surface. *Appl. Surf. Sci.* **2018**, *444*, 87–96. [CrossRef]
22. Chanturiya, V.; Masloboev, V.; Makarov, D.; Mazukhina, S.; Nesterov, D.; Men'shikov, Y. Artificial geochemical barriers for additional recovery of non-ferrous metals and reduction of ecological hazard from the mining industry waste. *J. Environ. Sci. Health Part A* **2011**, *46*, 1579–1587. [CrossRef] [PubMed]
23. Cao, Q.; Zou, H.; Chen, X.; Yu, X. Interaction of sulfuric acid with dolomite (104) surface and its impact on the adsorption of oleate anion: A DFT study. *Physicochem. Probl. Miner. Process.* **2020**, *56*, 34–42.
24. Yi, H.; Zhang, X.; Zhao, Y.; Liu, L.; Song, S. Molecular dynamics simulations of hydration shell on montmorillonite (001) in water. *Surf. Interface Anal.* **2016**, *48*, 976–980. [CrossRef]
25. Li, E.; Du, Z.; Yuan, S.; Cheng, F. Low temperature molecular dynamic simulation of water structure at sylvite crystal surface in saturated solution. *Miner. Eng.* **2015**, *83*, 53–58. [CrossRef]
26. Yi, H.; Jia, F.; Zhao, Y.; Wang, W.; Song, S.; Li, H.; Liu, C. Surface wettability of montmorillonite (0 0 1) surface as affected by surface charge and exchangeable cations: A molecular dynamic study. *Appl. Surf. Sci.* **2018**, *459*, 148–154. [CrossRef]

minerals

MDPI

Article

Synergistic Effect of Frequently Found Ions in the Flotation of Pb-Zn Sulfide Ores on Air/Water Interface

Can Gungoren [1], Saleban Mohamed Muse [1], Mert Terzi [1], Mehmet Faruk Eskibalci [1], Ilgin Kursun Unver [1] and Orhan Ozdemir [2,*]

[1] Mining Engineering Department, Engineering Faculty, Istanbul University-Cerrahpasa, 34500 Istanbul, Türkiye; can@iuc.edu.tr (C.G.); saleban454@gmail.com (S.M.M.); mert.terzi@iuc.edu.tr (M.T.); eskibal@iuc.edu.tr (M.F.E.); ilginkur@iuc.edu.tr (I.K.U.)

[2] Mineral Processing Engineering Department, Faculty of Mines, Istanbul Technical University, 34469 Istanbul, Türkiye

* Correspondence: orhanozdemir@itu.edu.tr

Abstract: The aqueous ions influence the properties of air bubbles and, therefore, the recovery of flotation. This study aims to reveal the synergistic effect of frequently found ions in the flotation of Pb-Zn sulfide ores. In this context, dynamic surface tension measurements, bubble coalescence time, Sauter mean diameter (SMD), bubble size distribution (BSD), and dynamic foam stability (DFS) measurements were carried out using artificial process waters (APWs). APW with the minimum ion concentration is expressed as "APW1" with the ionic strength (I) of 0.03 mol/dm^3. The concentration of the ions in APW1 was increased by 3, 5, and 10 times, and thus APW3 (I = 0.08 mol/dm^3), APW5 (I = 0.13 mol/dm^3), and APW10 (I = 0.26 mol/dm^3) were prepared, respectively. The results of this study indicated that the surface tension increased slightly in the presence of APW related to the ion concentration. Potassium ethyl xanthate (KEX) at high concentrations was effective in the reduction of surface tension. As the APW concentration increased, finer bubbles were obtained with a narrower size range. The stability of the foam increased with butyl glycol (BG) and APW concentration. There was no need to use a frother (BG) for the flotation experiments in the presence of APW1 or APWs with higher ionic strength.

Keywords: flotation; Pb-Zn; ions; APW; SMD; DFS

1. Introduction

The properties of air bubbles and the relation between them play a key role in flotation efficiency and selectivity. To obtain an acceptable recovery and concentration grade during flotation, the flotation bubbles must have a proper size distribution (bubble size distribution, BSD) considering the size of the particles to be floated [1]. The diameter of a group of spherical objects of different sizes, such as particles, droplets, or bubbles can be expressed as a single value, which is called the mean diameter. Sauter mean diameter (SMD, d_{32}) is one of the leading methods used to calculate the mean diameter [2,3].

The size of the air bubbles formed in flotation depends on the air flow rate as well as the characteristics of the air distribution system, such as geometry and pore size. However, bubbles formed in water are not stable. When two bubbles get too close to each other, they tend to coalesce to lower the interfacial area, and the total interfacial/surface energy. In this process, the liquid film between two bubbles thins and eventually ruptures [4–7]. As a result, the two bubbles coalesce to form a single larger bubble. This phenomenon is called bubble coalescence and adversely affects the flotation efficiency by making bubble–particle collision and attachment difficult [8] as well as causing the detachment of attached particles [9,10].

In flotation, the air–water interface tension, and hence bubble coalescence and, eventually, bubble size, is usually controlled by surfactants called frothers. Because of their amphiphilic character, the frother molecules are adsorbed on the air–water interface, thereby reducing the surface tension. Adsorbed surfactants including frothers decrease the surface tension of bubbles and thus decrease the overall surface energy of a system for the same surface area. Importantly, adsorbed surfactants also act as steric stabilizers, providing a barrier to the close contact of bubbles and altering the dynamics of the thin-film drainage process [11]. Thus, the coalescence of the bubbles is prevented and it is ensured that the bubbles remain at smaller sizes [12]. For this reason, the surface tension of a frother solution is an important indicator of the surface activity of the frother molecules.

Another structure in which bubble coalescence and bursting are controlled in flotation is the froth. The stability of the froth plays an important role in the flotation efficiency and the grade of concentrate [13,14]. Froth stability can be defined as the resistance of bubbles in the froth to coalescence. A more stable froth consists of smaller bubbles and less bubble coalescence and bursting occurs in the froth. Froth is a structure that consists of solids (mineral particles), liquid (water), and gas (air) phases, and the presence of mineral particles contributes to the stability of the froth. Efficient recovery of hydrophobic particles is not possible with an unstable froth. However, an extremely stable froth also increases the mechanical transport of hydrophilic particles, causing a decrease in the grade of the concentrate. Therefore, there is optimum froth stability for each flotation cell and condition in terms of flotation recovery and concentrate grade [15].

Meanwhile, in the absence of particles, this structure is called foam and the stability is provided by the surfactants (frothers). Although frothers are the main chemicals used to control bubble coalescence, and thus bubble size and foam stability, studies have shown that some dissolved ions also inhibit bubble coalescence above a concentration called the transition concentration [16].

Due to the scarcity of freshwater resources and increasing environmental sensitivity, low-quality water resources such as seawater and bore water can be used in industrial processes [17,18]. Another preferred method is the recycling and reuse of the water used in the plant [19]. However, there are many dissolved ions in these low-quality waters and the presence of these ions affects the size and the relation between bubbles by changing the properties of the air/water interface. While some of these ions affect the flotation process positively, some of them have a negative effect. Therefore, consistent and controllable water quality is essential to keep flotation efficiency at an acceptable level.

The effect of dissolved ions in flotation was mostly studied in the flotation of soluble salt minerals such as potash, trona, and borax [20–24]. However, there are also studies on nonsoluble ores [25–29] and coal [16].

The effect of aqueous ions on the air/water interface is ion-specific. Therefore, experiments are usually performed in the presence of a particular cation/anion pair using a single salt. However, the water used in industry contains many ions dissolved from the earth's crust, the ore, and the chemicals used in the beneficiation processes. Therefore, it is important to observe the synergistic effect of these ions with each other and with the flotation reagents. For this purpose, there are studies in the literature on water resources containing multiple ions and flotation reagents [30,31].

This study aimed to investigate the synergistic effect of frequently found ions in the flotation of Pb-Zn sulfide ores on the air/water interface. For this purpose, dynamic surface tension, bubble coalescence time, Sauter mean diameter (SMD), bubble size distribution (BSD), and dynamic foam stability (DFS) measurements were carried out using artificial process waters (APWs).

2. Materials and Methods

In this study, APWs were prepared to contain Na^+, Ca^{2+}, and Mg^{2+} cations, which are frequently encountered in industrial process waters, as well as Pb^{2-}, Zn^{2+}, Cu^{2+}, and Fe^{3+} metal cations found in flotation waters of sulfide Pb-Zn ores. The analytical-quality salts of these cations (NaCl (>99.5%), $CaCO_3$ (>98.0%), $MgSO_4 \cdot 7H_2O$ (>98.0%), $Pb(NO_3)_2$

(>98.0%), $ZnCl_2$ (>98.0%), $CuCl_2$ (>98.0%), and $Fe(NO_3)_3 \cdot 9H_2O$ (>99.0%)) were used to provide anion diversity. The salts were provided by Tekkim Chemical Company, Türkiye.

APW with the minimum ion concentration is expressed as "APW1" and the concentrations of Na^+, Ca^{2+}, and Mg^{2+} cations in APW1 were determined from the literature [32,33]. Meanwhile, the concentrations of Pb^{2+}, Zn^{2+}, Cu^{2+}, and Fe^{3+} cations were accepted as 1 ppm. The ion concentrations in APW1 were increased 3, 5, and 10 times, and thus APW3, APW5, and APW10 were prepared, respectively. The ion content of the APWs used in this study is given in Table 1.

Table 1. Ion types and concentrations in APWs (TDS: Total dissolved solids, I: Ionic strength).

	Na^+ (ppm)	Ca^{2+} (ppm)	Mg^{2+} (ppm)	Pb^{2+} (ppm)	Zn^{2+} (ppm)	Cu^{2+} (ppm)	Fe^{3+} (ppm)	$SO_4{}^{2-}$ (ppm)	Cl^- (ppm)	$CO_3{}^{2-}$ (ppm)	$NO_3{}^-$ (ppm)	TDS (mg/dm^3)	I (mol/dm^3)	pH
APW1	153	80	70	1	1	1	1	276.9	238.2	120	3.4	945.5	0.03	6.72
APW3	459	240	210	3	3	3	3	830.8	7146	360	10.1	2836.5	0.08	6.80
APW5	765	400	350	5	5	5	5	1384.7	1191	600	16.8	4727.5	0.13	6.80
APW10	1530	800	700	10	10	10	10	2769.4	2382	1200	33.0	9454.9	0.25	7.03

The pH of the flotation medium can reach up to 10–11 in Pb-Zn flotation facilities, and in this case, the ions in the metal medium precipitate in the form of $M_x(OH)_{y(s)}$. Meanwhile, the pH of the APWs was close to the pH of pure water, and therefore the experiments in this study were carried out at the natural pH of 6–7, as seen in Table 1. It is seen from the literature that at these pH values, most of this medium exists in the form of ions [34–40].

The glassware used in the experiments was cleaned with analytical ethanol (>99%), then rinsed with pure water (total dissolved solids (TDS) = 0–3 ppm), and finally dried in a drying oven (105 °C) before each use. In the experiments, potassium ethyl xanthate (KEX) (>96%, Sigma-Aldrich, St. Louis, MO, USA) and ethylene glycol mono butyl ether, also known as butyl glycol (BG), (>99.9% $C_4H_9OCH_2CH_2OH$, Tekkim, Türkiye), which are frequently preferred collectors and frothers, respectively, for the flotation of sulfide Pb-Zn ores, were used. The required solutions were freshly prepared before each experimental study via dispersion with a magnetic stirrer at a stirring speed of 500 rpm for 4 min at room temperature (23 ± 1 °C).

2.1. Dynamic Surface Tension Measurements

Dynamic surface tension measurements were carried out according to the pendant drop method using Attension Theta Lite (Biolin Scientific, Göteborg, Sweden) optical tensiometer. The volume of the droplets for pure water was 12.0 ± 0.5 cm^3 at 23 ± 1 °C. In this method, the changes in the shape of the drop at the tip of a backlit needle in the relevant solution were observed with the CCD (charge-coupled device) camera of the device. The surface tension of the solutions was calculated automatically by taking measurements at regular time intervals [30]. The measurements were started immediately after the formation of the drops and continued for 180 s. The surface tension values at 180 s were deemed as the equilibrium surface tension by observing the surface tension changes in time. The measurements were repeated three times and the average values were calculated.

2.2. Bubble Coalescence Time Measurements

In this study, bubble coalescence time measurements were performed with an induction time test unit (BKT-100, Bratton Engineering, Littleton, MA, USA) [41]. In the experiments, a glass container with a glass capillary tube was filled with the desired solution and placed under the glass capillary tube of the measurement device. Two bubbles of similar size (~2 mm) were formed at the ends of the glass tubes via a micro-syringe. The contact time was set in milliseconds via the computer software. When the device was started, the upper capillary tube moved downwards at 100 mm/s velocity and the two bubbles remained in contact for the specified time. Then, the upper capillary tube returned to its original position. The experimental setup and the sequential processes that occur in bubble coalescence time measurements are shown in Figure 1. The bubble coalescence experiments were carried out with twenty repetitions.

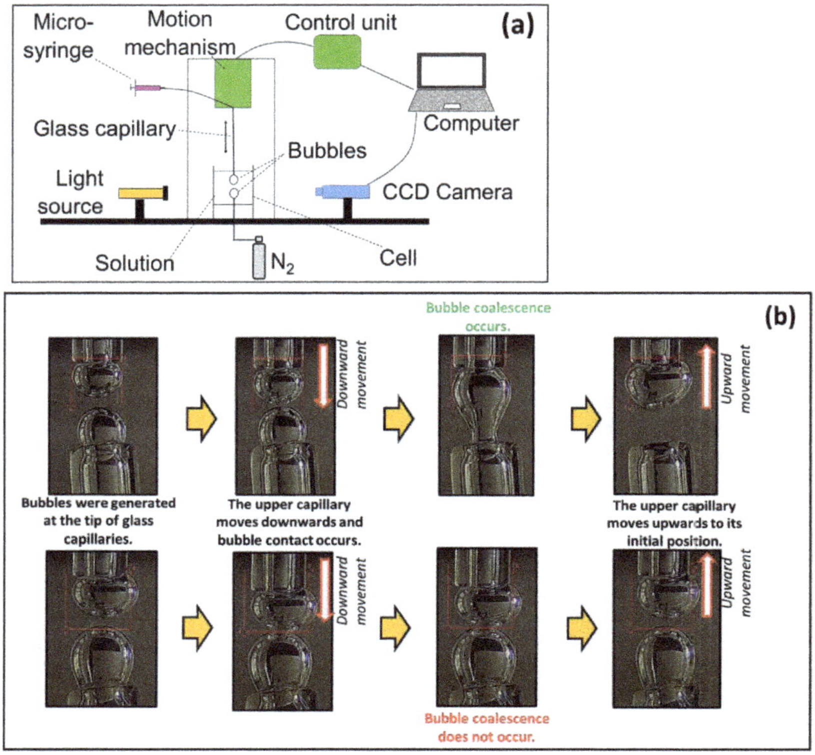

Figure 1. (**a**) Experimental setup for bubble coalescence time measurements and (**b**) the sequential processes that occur in bubble coalescence time measurements (The red dot lines is used for the adjustment of the distance between two bubbles during the measurements).

2.3. Measurements of Sauter Mean Diameter (SMD) and Bubbles Size Distribution (BSD)

SMD and BSD measurements were carried out using micro-flotation cells with a diameter of 30 mm, a length of 220 mm, and a volume of 155 cm^3. The effect of ion (APW) concentration on bubble size was investigated using various frit pore sizes (10–16, 16–40, and 40–100 µm) and nitrogen (N_2) at various flow rates (30 cm^3/min, 50 cm^3/min, and 70 cm^3/min).

In the measurements, the micro-flotation cell was filled with an aqueous solution of the desired chemical to the appropriate height, and representative images of the bubbles (photo and video) during the experiments were taken using a CCD camera (30 fps) with the appropriate lens (Bushman CMOS 8 Led Microscope, Bushman Equipment Inc., Menomonee Falls, WI, USA) at 50:1 magnification rate (Figure 2). For each concentration, at least ten images taken from the same position were analyzed. SMDs were determined using ImageJ open-source image processing software. The perimeter of the bubbles was measured automatically by the software. Using the perimeter values, the volume (V_b), the area (A_b), and thus the SMD (d_{32}) of the bubbles were calculated ($d_{32} = 6V_b/A_b$) [42].

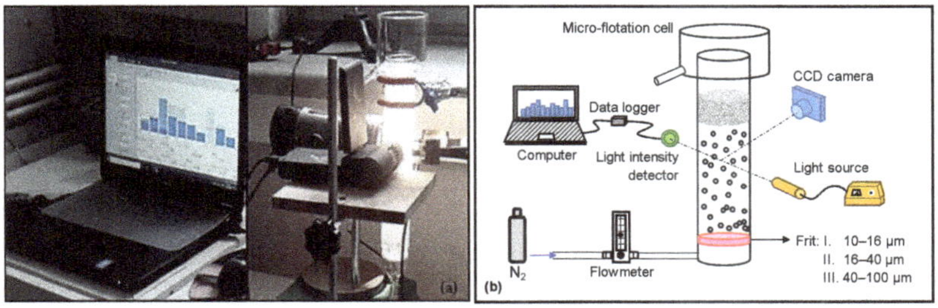

Figure 2. (**a**) Experimental setup for the SMD and BSD measurements as well as the determination of CCC and (**b**) the schematic view of the setup (The entrance of the N_2 is represented by the blue arrow, the bubbles in the system is represented by the blank circles, and the frit is represented with the red cylinder).

2.4. Determination of the Critical Coalescence Concentration (CCC) of Butyl Glycol

The critical coalescence concentration (CCC) is the minimum concentration of a frother that effectively prevents bubbles from coalescing. By determining the CCC value, the optimum concentration of a frother can be selected for a flotation process [43]. In the experiments, the micro-flotation cell was filled with the frother (BG) solution. Then, a light beam was sent through the cell using a cold light source (5100-L, SOIF, Shanghai, China). In this method, while some of the light passing through the column is absorbed by the solution, the intensity (mW) of the light passing through the column was measured by the light intensity detector (S121C, Thorlabs, Newton, NJ, USA) related to the coalescence of the bubbles in the presence of the frother at a constant concentration. Therefore, the CCC of the frother was determined. The experimental setup used in SMD and BSD measurements as well as for the determination of CCC is shown in Figure 2.

2.5. Dynamic Foam Stability (DFS) Measurements

The foaming ability of dissolved ions in the absence and presence of frother (BG) and their effect on DFS was investigated using a dynamic foam analyzer (DFA-100, Krüss, Hamburg, Germany). For the measurements, the required solutions (100 cm^3) were placed into the cylindrical glass measurement cell of the device (inner diameter: 40 mm, height: 250 mm). The foam formation is provided by dispersing the compressed air (200 cm^3/min, 250 cm^3/min, or 300 cm^3/min) through a filter plate located under the measurement cell.

The height of the solution and foam and, thus, their volumes were determined by the Krüss Advance Foam Analysis Software DFA100 by measuring the time-dependent light transmission utilizing an LED panel and a light detector positioned on the opposite sides of the measurement cell (Figure 3). At the end of the foaming period, the airflow was automatically stopped by the device. The height of the foam was measured for 60 s and the maximum foam height (H_{max}) was determined.

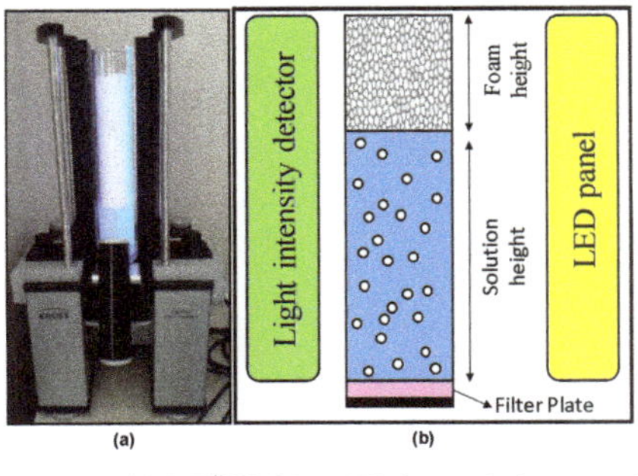

Figure 3. (**a**) KRÜSS DFA-100 dynamic foam analyzer and (**b**) the schematic diagram of the system (The bubbles in the system are represented by the blank circles).

3. Results and Discussion

3.1. Dynamic Surface Tension Measurements

The dynamic surface tensions of aqueous BG solutions (in pure water) were measured for 180 s. Since there was no significant change in surface tension at low BG concentrations, measurements were carried out at higher BG concentrations (1×10^{-5}, 1×10^{-4}, 1×10^{-3}, 1×10^{-2}, 1×10^{-1} mol/dm^3, and pure BG). The results of the measurements are given in Figure 4 with the equilibrium surface tension profile of BG solutions at 180 s. In Figure 4b, BG concentrations are given both in "ppm" and "mol/dm^3" units to compare the results with the other experiments carried out within this study.

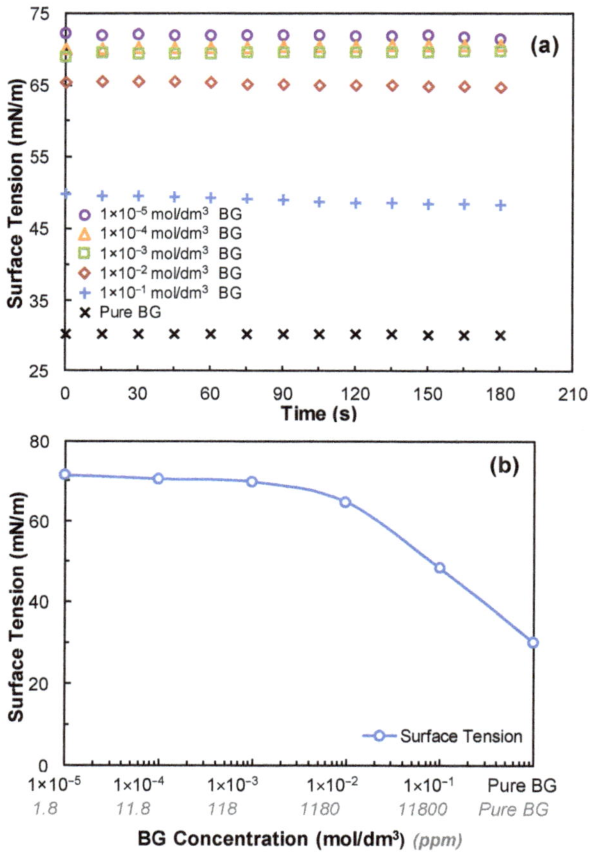

Figure 4. (**a**) Dynamic surface tension measurement results for BG at various concentrations and (**b**) equilibrium surface tension values at 180 s as a function of BG concentration in pure water.

Figure 4a shows that the surface tension of aqueous solutions of BG did not change significantly as a function of time. This shows that BG molecules reached the air/water interface immediately and the adsorption was completed in a short time. However, the surface tension decreased considerably in relation to the BG concentration. As seen in Figure 4b, while the surface tension is 71.3 mN/m at 1×10^{-5} M BG concentration, it decreased to 70.4 mN/m, 69.6 mN/m, 64.7 mN/m, and 48.3 at 1×10^{-4}, 1×10^{-3}, 1×10^{-2}, and 1×10^{-1} mol/dm^3 BG concentrations, respectively. Meanwhile, in the presence of pure BG, the surface tension was measured as 30.2 mN/m. This value is close to the results of previous studies by Smallwood [44] and Velasco-Medina and Gracia-Fadrique [45] (27.4–27.7 mN/m at 20 °C).

The dynamic surface tension measurement results with respect to KEX concentration and the equilibrium surface tension values at 180 s are given in Figure 5a,b, respectively. In the measurements, as there was no significant change was observed in surface tension in measurements made at dosages up to 100 g/t KEX, higher concentrations in mol/dm^3 were studied.

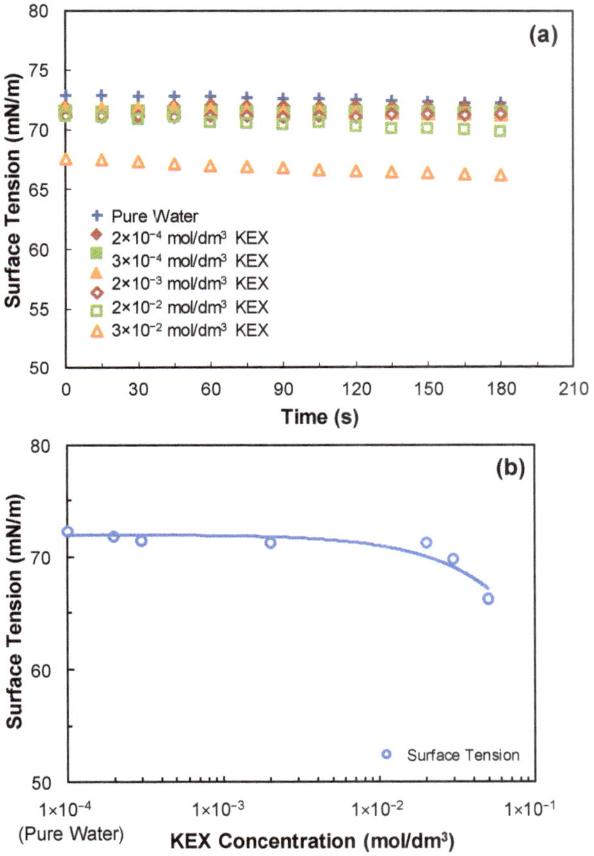

Figure 5. (**a**) Dynamic surface tension measurement results at various KEX concentrations and (**b**) equilibrium surface tension values at 180 s with respect to KEX concentration.

As seen in Figure 5a, the surface tension of aqueous KEX solutions showed a slight decrease over time at higher KEX concentrations. This shows that it takes time for KEX molecules to reach the air/water interface, especially at high concentrations. The selectivity of KEX pertains to the mineral/water interface. Furthermore, three carbon atoms exist in the hydrocarbon chain structure of KEX, which is not a sufficient chain length to rapidly lower the surface tension. However, when the KEX concentration increases to very high values, KEX molecules can penetrate the air/water interface and reduce the surface tension slightly. We obtained 66.1 mN/m surface tension at 5×10^{-2} mol/dm^3 KEX.

The results of dynamic surface tension measurements at various APW concentrations are given in Figure 6a and the equilibrium surface tension values at 180 s are shown in Figure 6b.

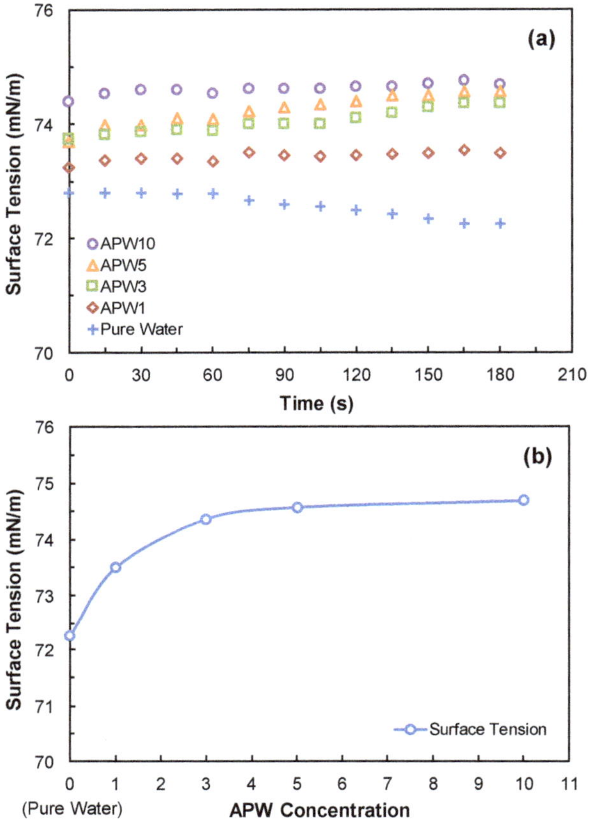

Figure 6. (**a**) Dynamic surface tension measurement results at various APW concentrations and (**b**) equilibrium surface tension values at 180 s with respect to APW concentration (in the absence of BG).

Figure 6a shows that the surface tension of APWs increased slightly with time. Meanwhile, the surface tension also increased with APW concentration, as seen in Figure 6b. While the surface tension is 73.5 mN/m in the presence of APW 1, it increased to 74.4 mN/m in the presence of APW3 and reached a plateau.

Several parameters are responsible for the change in the surface tension in the presence of ions, including the long-range electrostatic interactions, the valance, radius, and polarizability of the ion. These parameters affect the relationship of an ion with water molecules. Some ions are hydrated and remain in the bulk water phase, thus increasing the surface tension. These ions are called kosmotropic ions. On the other hand, chaotropic ions move to the air/water interface and decrease the surface tension. If there are surfactants adsorbed at the interface, the chaotropic ions penetrate them.

The chaotrophy/kosmotrophy property of an ion can be determined based on how long the water molecule is present near that ion. If a water molecule stays near an ion for longer on average than it does around another water molecule in bulk water, then this ion is called a kosmothropic ion. On the contrary, if a water molecule resides around an ion for a shorter average time than it resides around another water molecule in bulk water, it is called a chaothropic ion. Smaller monovalent ions usually have large electric fields on their surfaces and are kosmothropes [46–50].

Ions are ordered by Hofmeister [51] according to the strength of such relationships with water molecules. As is also known from the literature, dissolved ions generally increase the surface tension of a solution related to their concentration, except for several ions such as HCl, $HClO_4$, and HNO_3 [52,53]. In the studies conducted by Marcus [54] and Marcus [52], ionic surface tension changes (k_i) at ~25 °C of aqueous ions used in this study were reported except for Fe^{3+} and their data justify the ~2 mN/m increase in surface tension seen in Figure 5.

The results of the surface tension measurements in the presence of APW and BG are given in Figure 7.

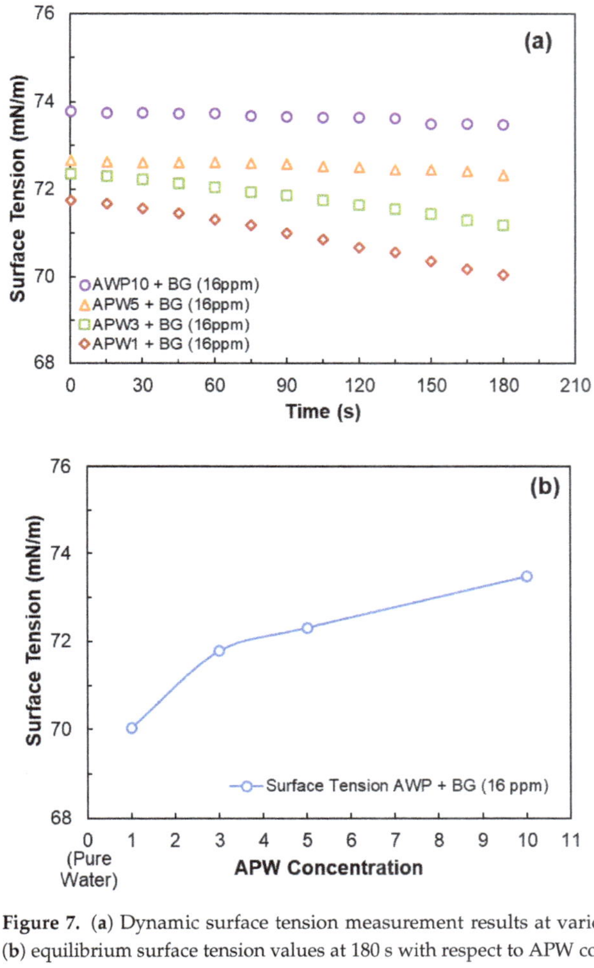

Figure 7. (**a**) Dynamic surface tension measurement results at various APW concentrations and (**b**) equilibrium surface tension values at 180 s with respect to APW concentration in the presence of BG (16 ppm).

In a system where both ions (APW) and BG are present at the same time, ions tend to increase the surface tension, whereas BG molecules tend to decrease it.

As seen in Figure 7a, although it takes time for BG molecules to reach the interface at low ion concentrations, BG can decrease the surface tension. In the presence of APW1 and BG at 16 ppm, the surface tension decreased from 71.7 mN/m to 70.0 mN/m after 180 s. However, this effect of BG decreased with the ion concentration. In the presence of APW10 and BG (16 ppm), the change in surface tension was only 0.5 mN/m.

Meanwhile, Figure 7b shows that surface tension increased with the APW concentration in the presence of BG (16 ppm). This indicates that ions are the dominant chemicals in the determination of surface tension.

3.2. Bubble Coalescence Time Measurements

Bubble coalescence percentage as a function of BG concentration is shown in Figure 8 for 1, 10, 100, and 1000 ms contact times.

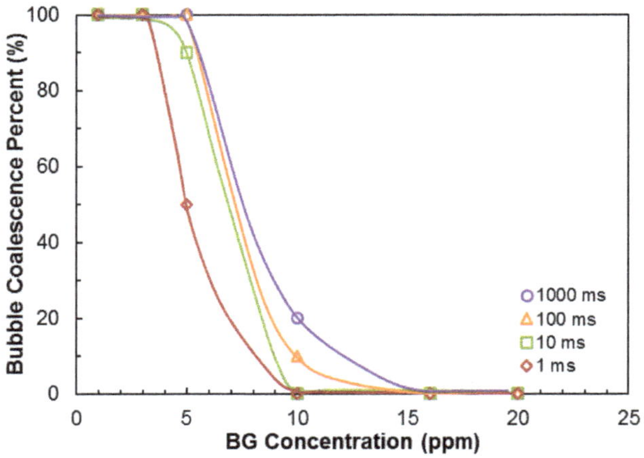

Figure 8. Bubble coalescence percentage with respect to BG concentration at 1, 10, 100, and 1000 ms contact times.

As seen in Figure 8, bubble coalescence could not be prevented up to ~5 ppm BG concentration. With the increase in the concentration of BG, the probability of bubble coalescence decreased, inversely proportional to the contact time. Higher BG concentrations were needed to prevent bubble coalescence at the same contact time. Bubble coalescence was completely inhibited in the presence of 16 ppm BG even at 1000 ms contact time. Therefore, it is clearly seen that bubble coalescence can be inhibited although there is no significant decrease in surface tension at these BG concentrations, as seen in Figure 4b.

In this study, the probability of bubble coalescence in the presence of APW was also investigated. The results indicated that bubble coalescence was completely inhibited (0% bubble coalescence was obtained) at all APW concentrations with or without BG and/or KEX in the medium. The bubble coalescence results of the experiments in the presence of APW with and without BG and KEX are given in Table 2. Meanwhile, Roberto and Arturo [55] stated that KEX decreased the coalescence time at alkaline pH conditions.

Table 2. Bubble coalescence results of the experiments in the presence of APW with and without BG and KEX.

Condition	Coalescence Result
APW1	No coalescence
APW3	No coalescence
APW5	No coalescence
APW10	No coalescence
APW1 + BG (16 ppm)	No coalescence
APW3 + BG (16 ppm)	No coalescence
APW5 + BG (16 ppm)	No coalescence
APW10 + BG (16 ppm)	No coalescence
APW1 + BG (16 ppm) + KEX (50 g/t)	No coalescence

Table 2. *Cont.*

Condition	Coalescence Result
APW3 + BG (16 ppm) + KEX (50 g/t)	No coalescence
APW5 + BG (16 ppm) + KEX (50 g/t)	No coalescence
APW10 + BG (16 ppm) + KEX (50 g/t)	No coalescence

3.3. SMD, BSD, and CCC Experiments

The effect of N_2 flow rate (30, 50, and 70 cm^3/min) on the SMD of the bubbles was revealed for all APW concentrations (APW1, 3, 5, and 10). As can be seen from the results given in Figure 9a, the lowest SMD values for APW1, 3, and 5 were obtained at 50 cm^3/min N_2 flow rate, while no significant change was observed in the presence of APW 10. Under these conditions, the optimum N_2 flow rate was accepted as 50 cm^3/min and was used in further experiments.

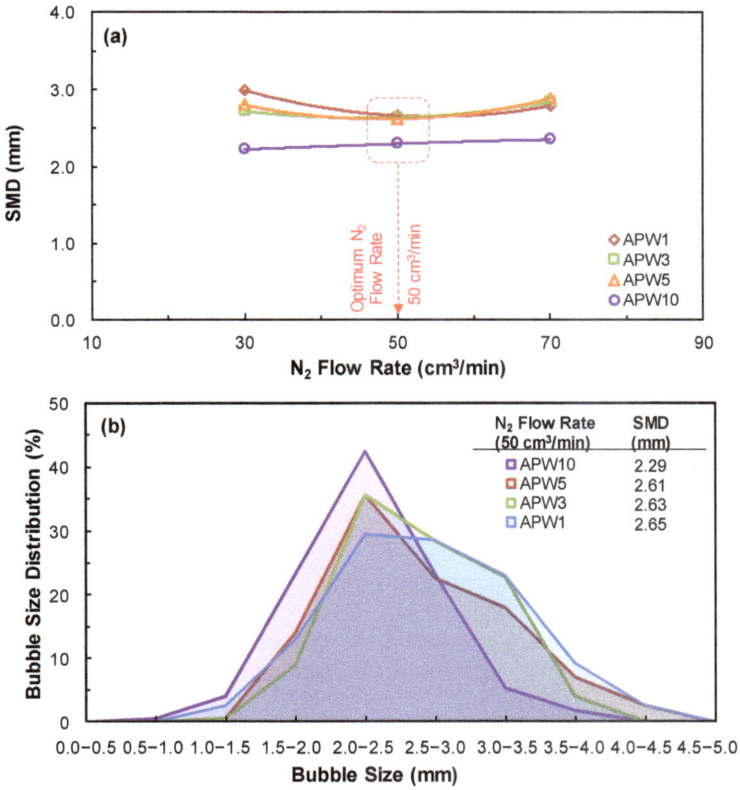

Figure 9. (**a**) SMD values of the bubbles as a function of N_2 flow rate for various APW concentrations and (**b**) bubble size distribution at various APW concentrations at 50 cm^3/min N_2 flow rate.

It is also seen in Figure 9a that while SMD decreased slightly with the ion concentration from APW1 to APW5, the decrease in the presence of APW10 was greater. As is known from the literature, the transition concentration of Na^+ is 1.5×10^{-5} mol/dm^3 [41]. Accordingly, the concentrations of some ions, such as Na^+, were higher than the transition concentrations in the presence of APW10, causing the SMD value to decrease more than other APW concentrations. The effect of physical parameters such as N_2 flow rate on SMD is relatively higher at low APW concentrations, but the effect of these physical parameters is minimized at high APW concentrations.

The bubble size distribution obtained at various APW concentrations at a constant N_2 flow rate of 50 cm^3/min is shown in Figure 9b. As seen in Figure 9b, the results of BSD for APW at various concentrations indicated that the bubbles in the presence of APW1 exhibit a wide size distribution. As the APW concentration increased, finer bubbles were obtained with a narrower size range.

The light intensity and bubble coalescence percentage depending on the BG concentration at a constant N_2 flow rate of 50 cm^3/min are given in Figure 10. It is seen from Figure 10 that the light intensity decreased as the BG concentration increased. The reason for this is that BG adsorbed to the air/water interface and reduced the surface tension so that the bubbles could remain stable at a smaller size without coalescing with other bubbles. As the number of bubbles in the system increased, the air/water interfaces formed by these bubbles refracted the light and, thus, the intensity of the light reaching the detector by passing across the measuring cell decreased. Bubble coalescence, which was 100% in the absence of BG, decreased with the BG concentration and became 50% in the presence of 16 ppm BG. The bubble coalescence was minimal after 100 ppm BG but this concentration was much higher than the BG concentrations used in Pb-Zn flotation.

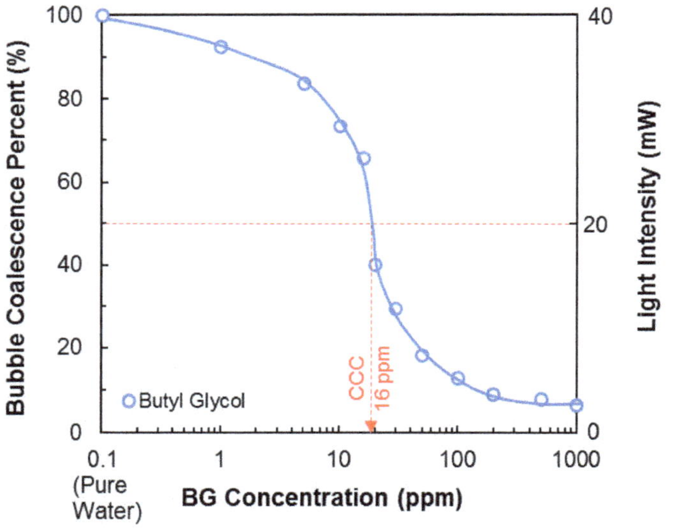

Figure 10. Light intensity and bubble coalescence percent values with respect to BG concentration at 50 cm^3/min N_2 flow rate (The 50% bubble coalescence percent and the CCC concentration is indicated with the red dotted line).

The change in the SMD of the bubbles obtained by using cells with various frit pore sizes (10–16, 16–40, and 40–100 μm) is given in Figure 11 as a function of APW concentration.

As seen in Figure 11, the SMD of bubbles in pure water produced using the frit with a pore size of 40–100 μm is 8.5 mm, while it is 6.2 and 5.7 mm at 16–40 μm and 10–16 μm pore sizes, respectively. SMD decreased with APW concentration for all pore sizes. Further studies were carried out using a cell with a frit of 10–16 μm, where the smallest bubble size was obtained. At this frit pore size, SMD reached a plateau at 2.7 mm in the presence of APW1.

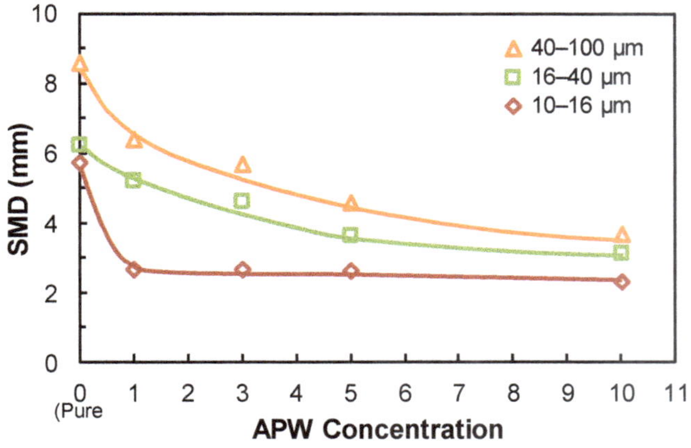

Figure 11. SMD values of the bubbles at various frit sizes as a function of APW concentration.

SMD values as a function of BG concentration at 50 cm^3/min N_2 flow rate are given in Figure 12.

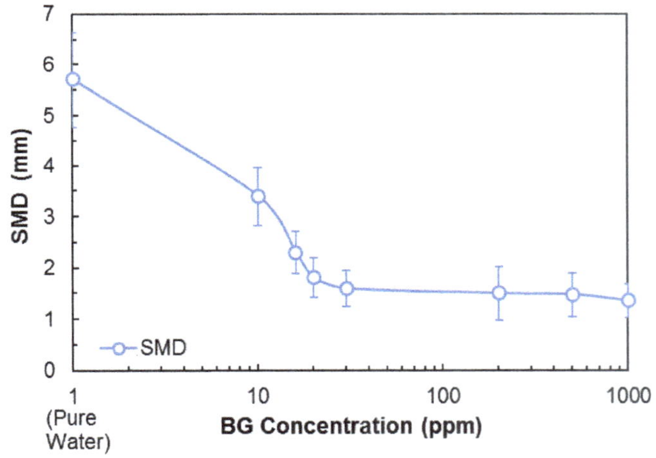

Figure 12. SMD values as a function of BG concentration at 50 cm^3/min N_2 flow rate.

As can be seen in Figure 12, while the SMD is 5.70 mm in the presence of pure water, it decreases with the addition of BG and reaches a plateau around 1.59 mm at 30 ppm BG. Kowalczuk, et al. [56] reported that CCC_{95} (the concentration giving a 95% reduction in bubble size compared to water only [57]) of BG is 0.236 $mmol/dm^3$, which is equal to 27.89 ppm.

SMD values at different APW concentrations are given in Figure 13.

As seen in Figure 13, the presence of dissolved ions in water in the absence of BG causes a rapid decrease in the SMD value. SMD, which is 5.70 mm in pure water, decreased to 2.65 mm in the presence of APW1 and reached a plateau. The SMD was 2.30 mm at 16 ppm BG (CCC value) in the absence of ions. A slight decrease was observed in the SMD in the presence of APW related to APW concentration and it reached 1.73 mm in the presence of APW10.

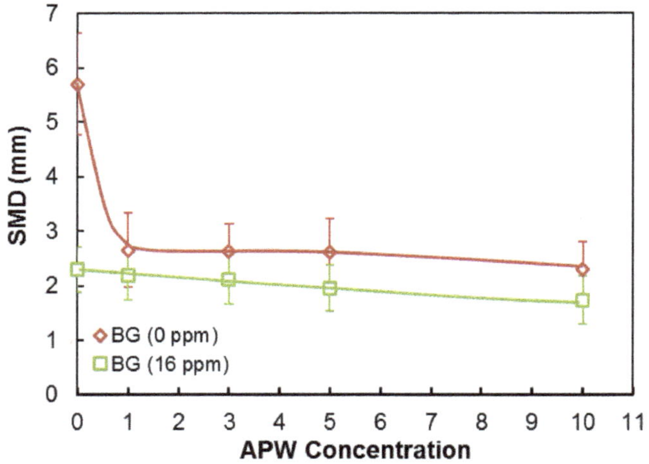

Figure 13. Effect of APW concentration on SMD in the absence of BG and the presence of 16 ppm (CCC) BG at 50 cm^3/min N_2 flow rate.

Meanwhile, dissolved ions can interact with flotation reagents such as frothers or collectors and have a synergistic effect. This synergistic effect may significantly influence the air/water interface [58,59]. Xu, et al. [60] reported that the presence of Na^+ in the medium decreased bubble size by reducing the surface charge of sodium dodecyl sulfate (SDS) micelles. Therefore, it promotes the adsorption of SDS molecules at the air/water interface. Yekeen, et al. [61] reported that surfactant (SDS) adsorption was higher in the presence of tri-(Al^{3+}) and di-valent (Ca^{2+}) ions compared to mono-valent ions (Na^+).

The bubble size distribution at 50 cm^3/min N_2 flow rate in the presence of APW and BG (16 ppm) is given in Figure 14, which indicates that bubbles with smaller size distribution were obtained in the presence of 16 ppm BG as the ion concentration increased.

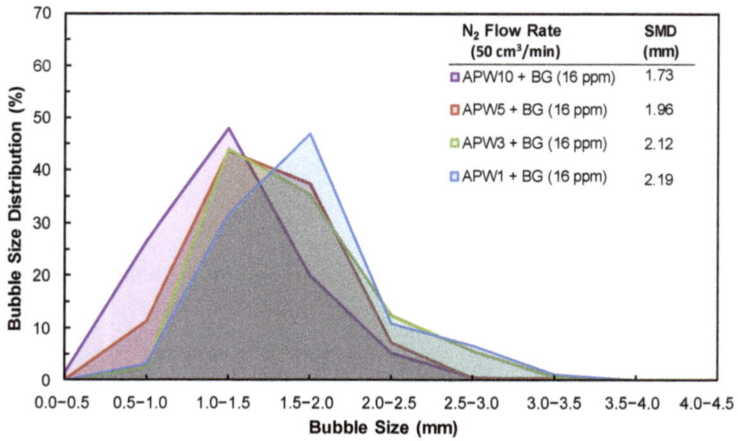

Figure 14. Bubble size distribution (50 cm^3/min N_2) at 16 ppm BG and various APW concentrations.

3.4. *Dynamic Foam Stability (DFS) Measurements*

The maximum foam heights (H_{max}) obtained in the DFS measurements are given as a function of N_2 flow rate, BG, and APW concentrations in Figure 15a–c, respectively.

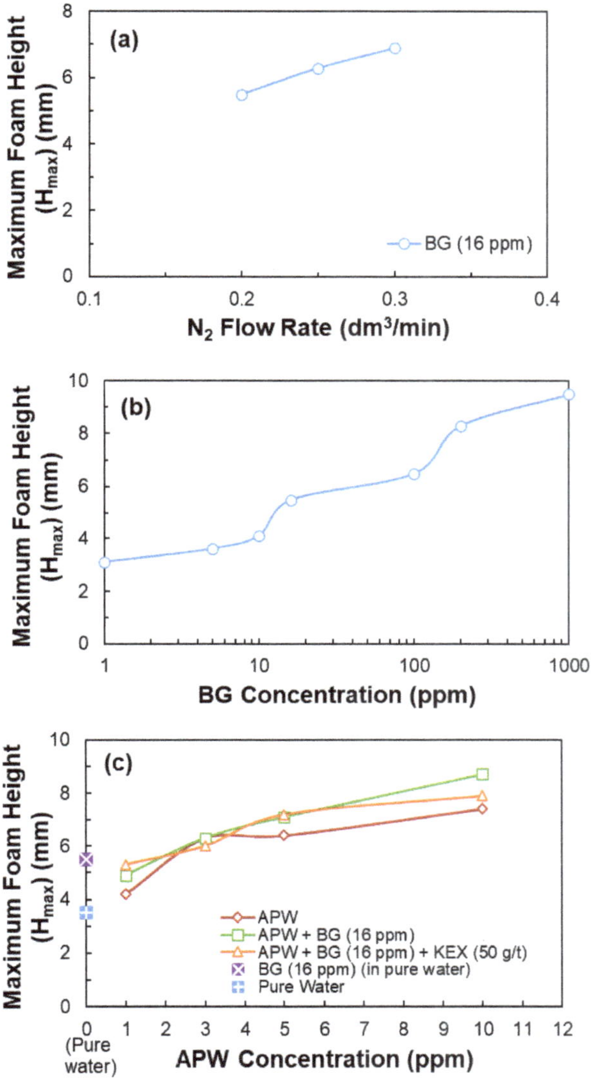

Figure 15. Maximum foam heights as a function of (**a**) N_2 flow rate in the presence of 16 ppm BG, (**b**) BG concentration at 0.2 dm^3/min N_2 flow rate, and (**c**) APW concentration.

Figure 15a indicates that when the air flow rate was 0.20 dm^3/min, the maximum foam height was 5.5 mm and it increased to 6.3 mm and 6.9 mm for 0.25 dm^3/min and 0.30 dm^3/min air flow rates, respectively. Further experiments were carried out at a 0.20 dm^3/min N_2 flow rate.

As seen in Figure 15b, the maximum foam height, which was 3.1 mm at 1 ppm BG concentration, increased to 5.5 mm in the presence of 16 ppm BG. At higher BG concentrations, the positive effect of BG on foam height remained moderate and a foam height of 9.5 mm was obtained in the presence of 1000 ppm BG.

It is seen in Figure 15c that the maximum foam height measured in the presence of pure water is 3.5 mm. This value can be accepted as the baseline. In this condition, since there is no chemical in the medium to ensure the stability of the foam, the bubbles formed will coalesce in a short time and burst. The foam height in the presence of APW1 was

4.2 mm. Considering that 5.5 mm foam height is obtained in the presence of 16 ppm BG (in pure water), it is seen that APW1 is quite beneficial for providing foam stability. The maximum foam height was further increased with the increase in APW concentration and reached 7.4 mm in the presence of APW10. Foam height was maximized when APW and BG were present together in the medium. A maximum foam height of 8.7 mm was reached at APW10 + BG (16 ppm). These results show that the presence of ions in the medium also increases the efficiency of BG.

The results of Yekeen, Manan, Idris and Samin [61] indicated that there is an optimum surfactant concentration, which maximizes the foam stability. This optimum concentration decreases with the ion concentration in the medium. Micelles starting to form at the critical micelle concentration (CMC) value of the surfactant increased foam drainage due to increasing weight. The main mechanism controlling the foaming and adsorption properties of surfactant is the screening effect of electrostatic double layer (EDL) by the ions and the formation of surfactant complex. The presence of ions promotes foam stability below the CMC with thicker lamellae.

It is seen from previous studies that ionic strength affects foam stability as well as frother dosage. It is seen that ionic strength increases the viscosity of the foam, slows down the liquid drainage between bubbles, decreases the bubble size, and increases the foam stability and foam decaying time [62,63]. This often increases the yield of valuable mineral recovery. However, a foam with very high stability will also increase water recovery and gangue transport into the concentrate, thereby reducing its grade [1,32,62].

Frother and xanthate influence the charge of the liquid film between bubbles in the foam related to their concentration. The cohesion in the surface layers and, thus, the stability of foam decreased with increasing xanthate concentration [64].

Meanwhile, the bubble coalescence experiment results of Pan, et al. [65] and Pan, et al. [66] showed that the presence of 1 ppm potassium amyl xanthate (KAX) considerably increased the lifetime of bubbles and the stability of foam compared to methyl isobutyl carbinol (MIBC)-only condition. According to Pan, et al. [67] the presence of NaCl enhanced the adsorption of frothers at the air/water interface, while the effect of KAX was insignificant.

4. Conclusions

In this study, the synergistic effect of frequently found ions in the flotation of Pb-Zn sulfide ores on the air/water interface was revealed with dynamic surface tension, bubble coalescence time, Sauter mean diameter (SMD), bubbles size distribution (BSD), and DFS measurements using artificial process waters (APWs).

The results of the surface tension measurements indicated that while the surface tension of water decreased with BG and KEX, the presence of ions in the APWs increased the surface tension as a function of their concentration.

As is well known from the literature, bubbles in the medium must remain stable at a certain size distribution without coalescing with each other for the bubble-hydrophobic particle attachment efficiency to be at an acceptable level in the flotation process. In flotation, this is made possible with the use of frothers (BG in the case of this study). As seen from the results of this study, bubble coalescence was completely inhibited in the presence of 16 ppm BG.

Additionally, the literature describes the significance of dissolved ions not only at mineral/air and mineral/water interfaces but also at air/water interfaces. For example, the dissolved ions behave like a frother in the system, showing that the bubble size in the pulp phase is reduced upon increasing the electrolyte concentration [68,69].

The results of this study showed that even in the absence of BG or KEX, the ion concentration in the presence of APW1 also completely inhibited bubble coalescence. The provided bubble stability continued at higher BG or ion (APW) concentrations. Moreover, the SMD of bubbles was 5.70 mm in pure water, which is a high size for flotation. This size decreased to a suitable size for flotation (2.30 mm) with the use of BG (16 ppm). When

APW1 was used in the absence of BG, the SMD of the bubbles grew close to the values in the presence of BG. Therefore, the presence of APW1 or APWs with higher ionic strength can provide suitable conditions for flotation in terms of bubble and froth stability even in the absence of a frother.

Author Contributions: Conceptualization, C.G., I.K.U. and O.O.; methodology, C.G. and O.O; investigation, S.M.M., C.G., M.T. and M.F.E.; writing—original draft preparation, C.G. and S.M.M.; writing—review and editing, C.G., O.O. and I.K.U.; visualization, C.G. and S.M.M.; supervision, I.K.U.; project administration, O.O.; funding acquisition, O.O. All authors have read and agreed to the published version of the manuscript.

Funding: This research was funded by The Scientific and Technological Research Council of Türkiye (TUBITAK), grant number 221M709.

Data Availability Statement: The original contributions presented in the study are included in the article.

Conflicts of Interest: The authors declare no conflict of interest. The funders had no role in the design of the study; in the collection, analyses, or interpretation of data; in the writing of the manuscript; or in the decision to publish the results.

References

1. Laskowski, J.S.; Cho, Y.S.; Ding, K. Effect of frothers on bubble size and foam stability in potash ore flotation systems. *Can. J. Chem. Eng.* **2003**, *81*, 63–69. [CrossRef]
2. Vazirizadeh, A.; Bouchard, J.; Chen, Y. Effect of particles on bubble size distribution and gas hold-up in column flotation. *Int. J. Miner. Process.* **2016**, *157*, 163–173. [CrossRef]
3. Kowalczuk, P.B.; Drzymala, J. Physical meaning of the Sauter mean diameter of spherical particulate matter. *Part. Sci. Technol.* **2016**, *34*, 645–647. [CrossRef]
4. Tao, D. Role of bubble size in flotation of coarse and fine particles—A review. *Sep. Sci. Technol.* **2005**, *39*, 741–760. [CrossRef]
5. Orvalho, S.; Ruzicka, M.C.; Olivieri, G.; Marzocchella, A. Bubble coalescence: Effect of bubble approach velocity and liquid viscosity. *Chem. Eng. Sci.* **2015**, *134*, 205–216. [CrossRef]
6. Ata, S.; Jameson, G.J. The formation of bubble clusters in flotation cells. *Int. J. Miner. Process.* **2005**, *76*, 123–139. [CrossRef]
7. Ata, S. Coalescence of bubbles covered by particles. *Langmuir* **2008**, *24*, 6085–6091. [CrossRef]
8. Zhu, H.; Valdivieso, A.L.; Zhu, J.; Song, S.; Min, F.; Corona Arroyo, M.A. A study of bubble size evolution in Jameson flotation cell. *Chem. Eng. Res. Des.* **2018**, *137*, 461–466. [CrossRef]
9. Ang, Z.; Bournival, G.; Ata, S. Influence of frothers on the detachment of galena particles from bubbles. *Int. J. Miner. Process.* **2013**, *121*, 59–64. [CrossRef]
10. Ata, S. The detachment of particles from coalescing bubble pairs. *J. Colloid Interface Sci.* **2009**, *338*, 558–565. [CrossRef] [PubMed]
11. Manor, O.; Chau, T.T.; Stevens, G.W.; Chan, D.Y.; Grieser, F.; Dagastine, R.R. Polymeric stabilized emulsions: Steric effects and deformation in soft systems. *Langmuir* **2012**, *28*, 4599–4604. [CrossRef] [PubMed]
12. Bournival, G.; Pugh, R.J.; Ata, S. Examination of NaCl and MIBC as bubble coalescence inhibitor in relation to froth flotation. *Miner. Eng.* **2012**, *25*, 47–53. [CrossRef]
13. Wang, B.; Peng, Y. The effect of saline water on mineral flotation–A critical review. *Miner. Eng.* **2014**, *66–68*, 13–24. [CrossRef]
14. Wei, T.; Peng, Y.; Farrokhpay, S. Froth stability of coal flotation in saline water. *Miner. Process. Extr. Metall.* **2014**, *123*, 234–240. [CrossRef]
15. Farrokhpay, S. The significance of froth stability in mineral flotation—A review. *Adv. Colloid Interface Sci.* **2011**, *166*, 1–7. [CrossRef] [PubMed]
16. Ozdemir, O. Specific ion effect of chloride salts on collectorless flotation of coal. *Physicochem. Probl. Miner. Process.* **2013**, *49*, 511–524. [CrossRef]
17. Liu, W.; Moran, C.J.; Vink, S. A review of the effect of water quality on flotation. *Miner. Eng.* **2013**, *53*, 91–100. [CrossRef]
18. Laskowski, J.; Castro, S. Flotation in concentrated electrolyte solutions. *Int. J. Miner. Process.* **2015**, *144*, 50–55. [CrossRef]
19. October, L.; Corin, K.; Schreithofer, N.; Manono, M.; Wiese, J. Water quality effects on bubble-particle attachment of pyrrhotite. *Miner. Eng.* **2019**, *131*, 230–236. [CrossRef]
20. Sun, K.; Nguyen, C.V.; Nguyen, N.N.; Ma, X.; Nguyen, A.V. Crucial roles of ion-specific effects in the flotation of water-soluble KCl and NaCl crystals with fatty acid salts. *J. Colloid Interface Sci.* **2023**, *636*, 413–424. [CrossRef]
21. Wu, Z.; Wang, X.; Liu, H.; Zhang, H.; Miller, J.D. Some physicochemical aspects of water-soluble mineral flotation. *Adv. Colloid Interface Sci.* **2016**, *235*, 190–200. [CrossRef] [PubMed]
22. Sun, K.; Nguyen, C.V.; Nguyen, N.N.; Nguyen, A.V. Flotation surface chemistry of water-soluble salt minerals: From experimental results to new perspectives. *Adv. Colloid Interface Sci.* **2022**, *309*, 102775. [CrossRef] [PubMed]

23. Weedon, D.; Grano, S.; Akroyd, T.; Goncalves, K.; Moura, R. Effects of high magnesium ion concentration on KCl flotation: Part II—Plant research. *Miner. Eng.* **2007**, *20*, 716–721. [CrossRef]
24. Weedon, D.; Grano, S.; Akroyd, T.; Goncalves, K.; Moura, R. Effects of high Mg^{2+} concentration on KCl flotation: Part I–Laboratory research. *Miner. Eng.* **2007**, *20*, 675–683. [CrossRef]
25. Choi, J.; Choi, S.Q.; Park, K.; Han, Y.; Kim, H. Flotation behaviour of malachite in mono- and di-valent salt solutions using sodium oleate as a collector. *Int. J. Miner. Process.* **2016**, *146*, 38–45. [CrossRef]
26. Jung, M.; Tadesse, B.; Dick, C.; Logan, A.; Dyer, L.; Albijanic, B. Influence of monovalent and divalent cations on monazite flotation. *Colloids Surf. A Physicochem. Eng.* **2022**, *653*, 129975. [CrossRef]
27. Wang, M.; Zhao, W.; Han, G.; Feng, Q. Utilization of lead ions to improve surface hydrophobicity and flotation recovery of sulfidized smithsonite. *Colloids Surf. A Physicochem. Eng.* **2023**, *663*, 131126. [CrossRef]
28. Manono, M.S.; Corin, K.C.; Wiese, J.G. An investigation into the effect of various ions and their ionic strength on the flotation performance of a platinum bearing ore from the Merensky reef. *Miner. Eng.* **2012**, *36–38*, 231–236. [CrossRef]
29. Wan, H.; Yi, P.; Song, X.; Luukkanen, S.; Qu, J.; Yang, W.; Li, H.; Bu, X. Role of improving molybdenite flotation by using aromatic hydrocarbon collector in high-calcium water: A multiscale investigation. *Miner. Eng.* **2023**, *191*, 107984. [CrossRef]
30. Gungoren, C. An investigation of air/water interface in mixed aqueous solutions of KCl, NaCl, and DAH. *Physicochem. Probl. Miner. Process.* **2019**, *55*, 1259–1270. [CrossRef]
31. Ozdemir, O.; Karakashev, S.I.; Nguyen, A.V.; Miller, J.D. Adsorption and surface tension analysis of concentrated alkali halide brine solutions. *Miner. Eng.* **2009**, *22*, 263–271. [CrossRef]
32. Manono, M.S.; Corin, K.C.; Wiese, J.G. The effect of ionic strength of plant water on foam stability: A 2-phase flotation study. *Miner. Eng.* **2013**, *40*, 42–47. [CrossRef]
33. October, L.L.; Manono, M.S.; Wiese, J.G.; Schreithofer, N.; Corin, K.C. Fundamental and flotation techniques assessing the effect of water quality on bubble-particle attachment of chalcopyrite and galena. *Miner. Eng.* **2021**, *167*, 1–7. [CrossRef]
34. Ornelas Tabares, J.; Madrid Ortega, I.; Reyes Bahena, J.L.; Sánchez López, A.A.; Valdez Pérez, D.; López Valdivieso, A. Surface properties and flotability of molybdenite. In Proceedings of the 2006 China-Mexico Workshop on Minerals Particle Technology, San Luis Potosí, Mexico, 5–7 December 2006; pp. 115–124.
35. Xue, J.; Fuchida, S.; Ishida, S.; Tokoro, C. Insight on Exogenous Calcium/Magnesium in Weakening Pyrite Floatability with Prolonged Pre-Oxidation: Localized and Concomitant Secondary Minerals and Their Depression Characteristics. *Minerals* **2022**, *12*, 115. [CrossRef]
36. Zhang, Q.; Zhu, H.; Yang, B.; Jia, F.; Yan, H.; Zeng, M.; Qu, H. Effect of Pb^{2+} on the flotation of molybdenite in the presence of sulfide ion. *Results Phys.* **2019**, *14*, 102361. [CrossRef]
37. Gao, Y.; Jiang, H.; Li, X.; Khoso, S.A.; Xiang, G.; Han, W. Different Insights into Silicate Rectorite Modification and Its Role in Removal of Heavy Metal Ions from Wastewater. *Minerals* **2020**, *10*, 176. [CrossRef]
38. Poliseno, V.; Chaves, S.; Brunetti, L.; Loiodice, F.; Carrieri, A.; Laghezza, A.; Tortorella, P.; Magalhaes, J.D.; Cardoso, S.M.; Santos, M.A.; et al. Derivatives of Tenuazonic Acid as Potential New Multi-Target Anti-Alzheimer's Disease Agents. *Biomolecules* **2021**, *11*, 111. [CrossRef]
39. Domínguez, S.; Torres, J.; Morales, P.; Heinzen, H.; Bertucci, A.; Kremer, C. Complexation and antioxidant activity of flavonoids with biologically relevant metal ions. In *Coordination Chemistry Research Progress*; Nova Science Publishers, Inc.: Hauppauge, NY, USA, 2008; pp. 305–328.
40. Yang, B.; Tong, X.; Deng, Z.; Lv, X. The Adsorption of Cu Species onto Pyrite Surface and Its Effect on Pyrite Flotation. *J. Chem.* **2016**, *2016*, 4627929. [CrossRef]
41. Gungoren, C.; Islek, E.; Baktarhan, Y.; Kurşun Unver, I.; Ozdemir, O. A novel technique to investigate the bubble coalescence in the presence of surfactant (MIBC) and electrolytes (NaCl and $CaCl_2$). *Physicochem. Probl. Miner. Process.* **2018**, *54*, 1215–1222. [CrossRef]
42. Cho, Y.S.; Laskowski, J.S. Bubble coalescence and its effect on dynamic foam stability. *Can. J. Chem. Eng.* **2002**, *80*, 299–305. [CrossRef]
43. Szyszka, D. Critical Coalescence Concentration (CCC) for Surfactants in Aqueous Solutions. *Minerals* **2018**, *8*, 431. [CrossRef]
44. Smallwood, I.M. Ethylene glycol monobutyl ether. In *Handbook of Organic Solvent Properties*; Smallwood, I.M., Ed.; Butterworth-Heinemann: Oxford, UK, 1996; pp. 131–133.
45. Velasco-Medina, A.A.; Gracia-Fadrique, J. Surface Tension and Density of Ethylene Glycol Monobutyl Ether–Water Mixtures from 277.15 to 308.15 K. *J. Chem. Eng. Data* **2020**, *65*, 4547–4555. [CrossRef]
46. Miller, D.J.; Lisy, J.M. Hydrated Alkali-Metal Cations: Infrared Spectroscopy and ab Initio Calculations of $M^+(H_2O)_{x=2-5}$Ar cluster ions for M) Li, Na, K, and Cs. *J. Am. Chem. Soc.* **2008**, *130*, 15381–15392. [CrossRef]
47. Huang, B.; Muy, S.; Feng, S.; Katayama, Y.; Lu, Y.C.; Chen, G.; Shao-Horn, Y. Non-covalent interactions in electrochemical reactions and implications in clean energy applications. *Phys. Chem. Chem. Phys.* **2018**, *20*, 15680–15686. [CrossRef] [PubMed]
48. Franks, G.V.; Johnson, S.B.; Scales, P.J.; Boger, D.V.; Healy, T.W. Ion-Specific Strength of Attractive Particle Networks. *Lungmuir* **1999**, *15*, 4411–4420. [CrossRef]
49. Hishida, M.; Kaneko, Y.; Okuno, M.; Yamamura, Y.; Ishibashi, T.A.; Saito, K. Communication: Salt-induced water orientation at a surface of non-ionic surfactant in relation to a mechanism of Hofmeister effect. *J. Chem. Phys.* **2015**, *142*, 171101. [CrossRef]

50. Hervo-Hansen, S.; Heyda, J.; Lund, M.; Matubayasi, N. Anion-cation contrast of small molecule solvation in salt solutions. *Phys. Chem. Chem. Phys.* **2022**, *24*, 3238–3249. [CrossRef]
51. Hofmeister, F. Zur Lehre von der Wirkung der Salze. *Arch. Exp. Path. Pharm.* **1888**, *24*, 247–260. [CrossRef]
52. Marcus, Y. Specific ion effects on the surface tension and surface potential of aqueous electrolytes. *Curr. Opin. Colloid Interface Sci.* **2016**, *23*, 94–99. [CrossRef]
53. Witecki, K.; Polowczyk, I.; Kowalczuk, P.B. Chemistry of wastewater circuits in mineral processing industry—A review. *J. Water Process. Eng.* **2022**, *45*, 102509. [CrossRef]
54. Marcus, Y. Individual ionic surface tension increments in aqueous solutions. *Langmuir* **2013**, *29*, 2881–2888. [CrossRef]
55. Roberto, P.-G.; Arturo, B.-T. Coalescence of air bubbles: Effect of the electrical double layer. *Miner. Eng.* **2020**, *150*, 106301. [CrossRef]
56. Kowalczuk, P.B.; Zaleska, E.; Danczak, O. Flotation of carbonaceous copper shale–quartz mixture with poly(ethylene glycol) alkyl ethers. *Trans. Nonferrous Met. Soc. China* **2015**, *25*, 314–318. [CrossRef]
57. Zhang, W.; Nesset, J.E.; Rao, R.; Finch, J.A. Characterizing frothers through critical coalescence concentration (CCC) 95-hydrophile-lipophile balance (HLB) relationship. *Minerals* **2012**, *2*, 208–227. [CrossRef]
58. Góralczyk, D.; Hąc, K.; Wydro, P. Surface properties of the binary mixed systems of alkylpyridinium halides and sodium alkylsulfonates. *Colloids Surf. A Physicochem. Eng.* **2003**, *220*, 55–60. [CrossRef]
59. Naik, P.K.; Reddy, P.S.R.; Misra, V.N. Interpretation of interaction effects and optimization of reagent dosages for fine coal flotation. *Int. J. Miner. Process.* **2005**, *75*, 83–90. [CrossRef]
60. Xu, Q.; Nakajima, M.; Ichikawa, S.; Nakamura, N.; Roy, P.; Okadome, H.; Shiina, T. Effects of surfactant and electrolyte concentrations on bubble formation and stabilization. *J. Colloid Interface Sci.* **2009**, *332*, 208–214. [CrossRef]
61. Yekeen, N.; Manan, M.A.; Idris, A.K.; Samin, A.M. Influence of surfactant and electrolyte concentrations on surfactant adsorption and foaming characteristics. *J. Pet. Sci. Eng.* **2017**, *149*, 612–622. [CrossRef]
62. Corin, K.C.; Tetlow, S.; Manono, M.S. Considering the action of frothers under degrading water quality. *Miner. Eng.* **2022**, *181*, 107546. [CrossRef]
63. Corin, K.C.; Reddy, A.; Miyen, L.; Wiese, J.G.; Harris, P.J. The effect of ionic strength of plant water on valuable mineral and gangue recovery in a platinum bearing ore from the Merensky reef. *Miner. Eng.* **2011**, *24*, 131–137. [CrossRef]
64. Manev, E.D.; Pugh, R.J. Drainage and equilibrium thickness of aqueous films containing nonionic frothers and xanthate flotation collector. *J. Colloid Interface Sci.* **1992**, *151*, 505–516. [CrossRef]
65. Pan, Y.; Bournival, G.; Ata, S. The role of non-frothing reagents on bubble size and bubble stability. *Miner. Eng.* **2021**, *161*, 106652. [CrossRef]
66. Pan, Y.; Bournival, G.; Karakashev, S.I.; Yang, B.; Wang, L.; Ata, S. Effect of electrolyte and flotation collector on foam films in the presence of frothers. *Chem. Eng. Sci.* **2022**, *264*, 118176. [CrossRef]
67. Pan, Y.; Gresham, I.; Bournival, G.; Prescott, S.; Ata, S. Synergistic effects of frothers, collector and salt on bubble stability. *Powder Technol.* **2022**, *397*, 117028. [CrossRef]
68. Manono, M.; Corin, K. Considering Specific Ion Effects on Froth Stability in Sulfidic Cu-Ni-PGM Ore Flotation. *Minerals* **2022**, *12*, 321. [CrossRef]
69. Wang, J.; Forbes, G.; Forbes, E. Frother Characterization Using a Novel Bubble Size Measurement Technique. *Appl. Sci.* **2022**, *12*, 750. [CrossRef]

minerals

Review

Review of Interfacial Regulation of Apatite Flotation

Zhe Liu [1], Lixia Li [1,*], Zhuguo Li [2], Meng Wang [2], Feifei Liu [1] and Hongcheng Mi [1]

1 College of Civil and Resources Engineering, Northeastern University, Shenyang 110819, China; 2210462@stu.neu.edu.cn (Z.L.)
2 Chengde Jingcheng Mining Group Co., Ltd., Chengde 067600, China
* Correspondence: lilixia@mail.neu.edu.cn; Tel.: +86-13842021890

Abstract: Phosphate ores, which are regarded as critical mineral resources, play an important role in various industrial fields. Apatite is the main source of phosphate mineral resources and must be concentrated before it is processed into industrial products. Flotation is the most commonly employed method for apatite concentration. However, as the proportion of fine apatite increases, the challenge of separating it from gangue minerals intensifies, due to the resemblance in surface characteristics between apatite and gangue. Interfacial regulation during flotation is fundamental to the process, including the regulation of the mineral/water interface wettability by flotation reagents (collectors and modifiers), the control of interactions between mineral particles, and the regulation of interactions between mineral particles and bubbles. This article introduces the surface characteristics of apatite and its main gangue minerals. It discusses innovative work on flotation reagents (primarily collectors and depressants) and their action mechanisms on mineral surfaces. It reviews the current development of theories on the regulation of interactions between interparticles and between particles and bubbles. Finally, the study outlook the future research on interfacial regulation in apatite flotation. This study is intended to offer references for the continued advancement of apatite flotation.

Keywords: apatite; interfacial regulation; reagents; interfacial chemistry; surface adsorption; fine particles

1. Introduction

Phosphate ores have been classified as critical mineral resources by several countries and regions, including China, the European Union, and the United States. Data have shown that the global phosphate ore reserves total 74 billion tons, with the majority being located in Africa [1]. Morocco holds the largest phosphate reserves, amounting to 50 billion tons, which accounts for 67.57% of the global total. China follows with reserves of 3.8 billion tons. In 2023, the global production of phosphate ore reached 2.2 billion tons, with China leading the world in production at 900 million tons. Morocco and the United States followed [1]. Phosphorus is extensively consumed in agriculture and is considered irreplaceable and non-recyclable. It is essential for ensuring food security, fostering economic growth, and advancing societal well-being [1–4]. Figure 1 displays the global allocation of phosphorus reserves, the varieties of phosphorus deposits, and the applications of phosphorus ores. Among them, collophane is characterized by low grade, high levels of harmful impurities, and poor consistency. Apatite, with its good flotation properties, has emerged as a vital resource for phosphorus extraction and the manufacturing of agricultural phosphate fertilizers, making it crucial in the processing of phosphate minerals [5–7]. Recently, the phosphate import and export trade has been

https://doi.org/10.3390/min15060558

impacted by international factors, heightening the urgency for countries to develop and utilize phosphate ore resources [8].

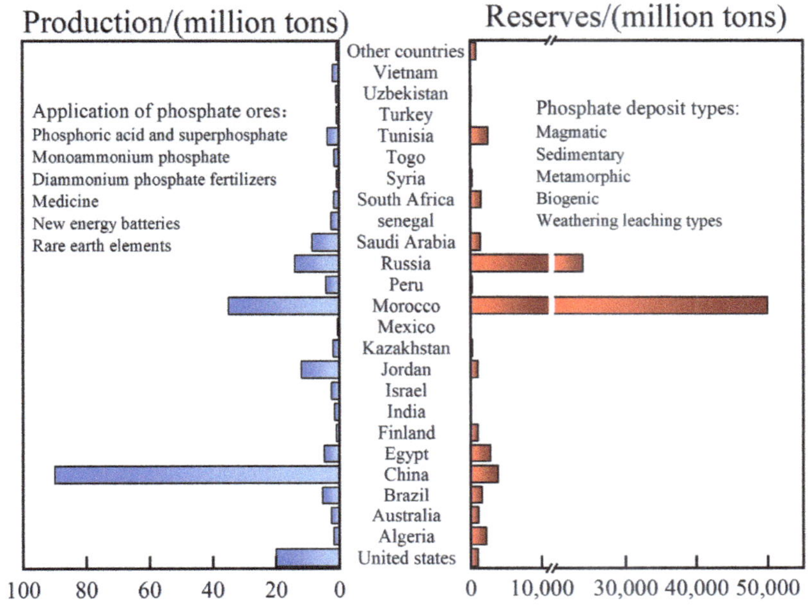

Figure 1. Distribution and application of world phosphorus resources [9–13].

In addition to phosphate minerals, phosphate ores also include multiple accompanying gangue minerals, including clay, quartz, carbonate-based minerals (primarily calcite and dolomite), organic material, iron oxides, and pyrite [14]. Depending on the chemical composition and content of the gangue minerals in the ore, apatite can be classified into siliceous, calcareous, or a mixture of both [15]. The beneficiation techniques for apatite ores vary depending on the specific apatite variety and associated gangue minerals. Apatite is non-magnetic and has a density similar to that of the gangue minerals; so, gravity and magnetic separation methods are seldom applied for its enrichment, while flotation remains the predominant technique for apatite beneficiation [16,17]. The basic principle of mineral separation through froth flotation is to control the surface properties and hydrophobicity of the ore particles using surfactants, thereby amplifying the differences in mineral floatability [18–20]. During flotation, the equipment produces air bubbles that preferentially adhere to hydrophobic mineral particles, carrying them upward to the pulp surface, whereas hydrophilic particles stay suspended in the slurry [21]. This differential bubble adhesion capitalizes on variations in the mineral surface's hydrophobicity and hydrophilicity, enabling effective apatite concentration [22]. It can be seen that factors such as particle diameter, contact angle, and bubble diameter significantly influence the effectiveness of flotation in this process [23].

Flotation was initially applied to the beneficiation of siliceous ores. Technological progress and enhanced knowledge of flotation chemistry and surface characteristics have enabled its application in processing both siliceous–carbonaceous and carbonaceous apatite ores [14,24,25]. In the context of the flotation enrichment of apatite, the most challenging aspect is the separation from gangue minerals such as quartz (SiO_2), calcite, and dolomite, particularly the separation from calcium-containing minerals like calcite and dolomite. The presence of calcium ions with comparable structural configurations in these minerals leads to similar surface chemical behaviors and dissolution patterns within the slurry medium.

This phenomenon limits the selectivity of adsorption of flotation reagents on surfaces containing calcium-rich minerals. Consequently, the separation process through flotation becomes challenging [26–29]. In addition, the ongoing development and utilization of mineral resources has resulted in the gradual depletion of high-grade phosphate deposits, leading to a decline in ore grades and a decrease in grain size. This necessitates the implementation of finer grinding techniques for monomer dissociation, thereby generating a substantial number of fine apatite particles [30–32]. Fine apatite particles (-20 μm) are characterized by their small size and low kinetic energy, which decreases the likelihood of contact and attachment with conventional flotation bubbles. Consequently, this leads to a diminished separation performance during the flotation process. Additionally, fine mineral particles possessing increased surface area demand higher dosages of flotation chemicals for effective treatment and frequently encounter challenges, including higher foam viscosity and mechanical entrapment of gangue minerals [33–35]. Furthermore, the presence of high levels of impurities in low-grade phosphate ores results in the complicated separation of apatite, reduced recovery rates at beneficiation plants, and increased separation costs [36].

Flotation represents a multiphase separation process encompassing solid particles, liquid media, and gas bubbles, governed by numerous chemical interactions and physical parameters. Chemical factors regulate the wettability of mineral surfaces through flotation reagents, while physical factors include ore properties, such as particle size and degree of liberation, as well as equipment-related factors such as aeration rate and bubble size [37]. Interfacial regulation in mineral flotation forms the foundation for addressing flotation challenges and mainly includes: (1) regulation of particle interfacial properties, primarily wettability and surface charge, and (2) regulation of interfacial interactions, including interactions between solid particles and the liquid phase (i.e., adsorption of reagents on mineral surfaces), interactions between solid particles (dispersion and aggregation), and interactions between solid particles and bubbles. It is, thus, imperative to conduct a comprehensive review of the key issues on the flotation process of apatite, incorporating a detailed analysis of its physicochemical properties, the reagents employed, the reagent reaction mechanisms, and interactions interparticles and between particles and bubbles. The present paper thus summarizes the crystal structure characteristics and surface physicochemical properties of apatite. A systematic review of the research progress on apatite flotation reagents, including collectors and depressants, is then conducted, followed by an analysis of the mechanisms of action of different reagent types. Furthermore, it summarizes the research advancements in understanding interparticles and particle–bubble interactions during apatite flotation. The overarching objective of this analysis is twofold: first, to comprehensively examine the challenges and underlying driving forces that govern the apatite flotation process; and second, to critically appraise the extant research conducted by the scientific community on these matters. The final aim is to propose future avenues for exploration in the field of apatite flotation separation theory.

2. Basic Properties of Apatite

The crystal-chemical principles of mineral flotation are predicated on the relationship between the mineral's crystal structure, chemical composition, surface properties, and floatability. A comprehensive understanding of the crystal chemistry and surface characteristics of minerals is essential for elucidating the mechanisms of mineral flotation.

2.1. Crystal Structure Characteristics of Apatite

The Commission on New Minerals, Nomenclature and Classification of the International Mineralogical Association (IMA-CNMNC) has determined the classification of apatite into three distinct types based on the predominant anions present (Cl^-, F^-, and OH^-).

This classification system identifies chlorapatite ($Ca_5(PO_4)_3Cl$), fluorapatite ($Ca_5(PO_4)_3F$), and hydroxylapatite ($Ca_5(PO_4)_3OH$) as the three primary types of apatite [38]. Fluorapatite is the most stable and prevalent form of apatite mineral [39]. Apatite, a mineral that is commonly found in nature, is known to undergo isomorphous substitution. Ideal pure apatite is seldom encountered [40]. The components fluorine (F), chlorine (Cl), and hydrogen oxide (OH) exhibit mutual substitutability, enabling complete solid solution formation. Within sedimentary phosphate deposits, the replacement of PO_4^{3-} by CO_3^{2-} and OH^- groups can result in the formation of carbonate-fluorapatite. This substitution form is the most prevalent in sedimentary deposits and is commonly designated as francolite [41].

The most predominant form of apatite mineral is fluorapatite. According to the American Mineralogist Crystal Structure Database (AMCSD), the ideal fluorapatite crystal has a complex hexagonal prismatic structure, with each unit cell containing 42 atoms. As shown in Figure 2, within fluorapatite crystalline structures, two distinct calcium position variants exist: Ca1 and Ca2. The Ca1 site is located between two layers of phosphate groups and is bonded to nine surrounding oxygen atoms. The Ca2 site is positioned within the same layer as the phosphate group. In this configuration, the Ca2 site is bonded to six adjacent oxygen atoms and a single fluorine atom [39]. Within the unit cell, two sets of Ca1 atoms (with two atoms per set) form atomic columns perpendicular to the (001) plane, while two sets of Ca2 atoms (with three atoms per set) form end faces perpendicular to the c-axis [42]. Research has indicated that the Ca1 site is more active and more likely to form unsaturated bond fractures. The fracture surfaces of apatite have been observed to exhibit two distinct types of bond fracture: Ca-O and Ca-F. Furthermore, extensive substitution has been observed in natural apatite crystals, including alkali metals (K and Na), transition metal ions (Mn, Ni, Cu, Co, and Zn), alkaline earth metals (Sr and Ba), and rare-earth elements at the Ca sites [42]. The (001) crystal face of apatite is the most common cleavage plane. This plane primarily exposes calcium (Ca) atoms and phosphate (PO_4^{3-}) groups in the sub-layer, with the outermost layer exposing only calcium (Ca) atoms. This facilitates the adsorption of flotation reagents [43]. Natural apatite crystals exhibit extensive substitution, including alkali metals (K and Na), transition metal ions (Mn, Ni, Cu, Co, and Zn), alkaline earth metals (Sr and Ba), along with rare-earth elements occupying calcium lattice positions [42].

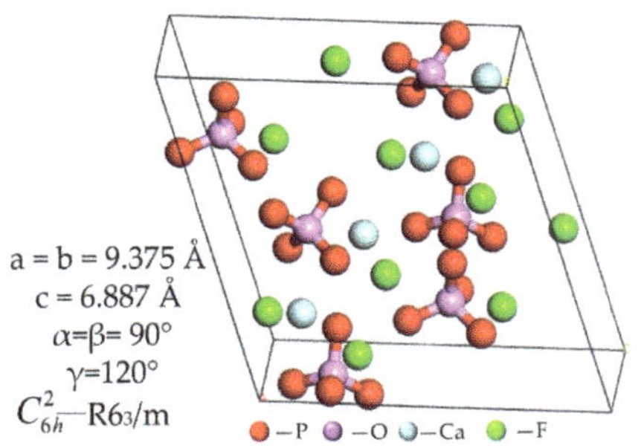

Figure 2. Fluorapatite crystal structure (a, b, c, α, β, γ: cell parameters; $C_{6h}{}^2$—$R6_3/m$: space group, C_{6h}: a six-fold rotational axis, and a horizontal mirror plane perpendicular to the six-fold rotational axis; R: rhombus; m: a mirror plane symmetry element) [44,45].

2.2. Surface Properties

Most flotation processes depend on the targeted attachment of collectors (organic surface-active agents) to mineral surfaces in aqueous environments. Detailed knowledge of mineral interfacial characteristics is essential for optimizing reagent formulations and improving separation efficiency. Measures of solid/liquid interface conditions, including contact angle measurements and zeta potential values, directly correlate with interfacial behaviors that manifest the intricate interactions occurring at the solid–liquid–gas boundary [46].

2.2.1. Surface Wettability

The hydrophobicity or hydrophilicity of mineral surfaces is quantified through contact angle analysis. The measurement of this parameter can be conducted employing multiple techniques, including the sessile drop approach, captive bubble analysis, and Washburn capillary rise (WCR) methodology. The regulation of mineral surface wettability by flotation reagents (mainly collectors) is the foundation of flotation separation. The contact angle of natural apatite surfaces is approximately 10°. However, due to the substitution of various ions and the deposition and impregnation of organic matter and impurities in the complex sedimentary environment, the contact angle on the surface of fluorapatite in sedimentary phosphate ores increases to some extent. The contact angle of apatite surfaces in sedimentary phosphate ores ranges from 50° to 66° [47]. Calcite and dolomite represent the dominant carbonate gangue in apatite deposits, while quartz serves as the principal silicate gangue mineral coexisting with apatite [48]. Since apatite, calcite, and dolomite share the presence of Ca^{2+} ions on their surfaces, as well as the deposition of calcium ions on the surface of quartz, they exhibit similar surface properties [32,49,50]. Through adjustments to the interfacial reactions between flotation chemicals and mineral surfaces, the flotation characteristics of minerals can be modified, generating significant variations in surface wettability that immediately influence separation effectiveness in flotation operations [51]. Table 1 presents measured contact angle values for apatite surfaces following treatment with various chemical reagents.

Table 1. Contact angle reported for apatite conditioned with some collectors.

Collector	Concentration	pH *	Angle (°)	Ref.	Year
NaOL	5×10^{-4} mol/L	Natural	96.7	[31]	2023
Mustard oil	450 mg/L	Natural	90	[52]	2024
Dodecylamine	20 mg/L	6.6	72.34	[53]	2025
Flaxseed oil	100 mg/L	Natural	86.8	[26]	2022
Nigella oil	300 mg/L	Natural	88.20	[26]	2022
TOFA-OAPEGE6 7:3	20 mg/L	9.5	88.03	[36]	2025
HDMEA	20 mg/L	Unknown	70.39	[54]	2022

*: The initial pH of the test.

Additionally, studies have shown that changes in the mineral surface roughness caused by grinding, as well as the dissolution of metal ions, are also important factors influencing wettability [55,56]. Some studies demonstrated that, for surfaces with a hydrophilic character, an increase in surface roughness resulted in an enhancement in hydrophilicity. Conversely, for surfaces characterized by hydrophobic properties, an increase in surface roughness led to an augmentation of hydrophobicity [57]. Jiang et al. [58] utilized density functional theory calculations to demonstrate that fluorapatite crystals containing impurities, such as Mg, Al, and Fe, manifest novel impurity energy levels. Moreover, these foreign elements induce a displacement of fluorapatite's density of states distribution relative to the Fermi energy level, thereby enhancing the hydrophobic character of the phosphate mineral surface. Observations have been made of the reaction between $H_2PO_4^-$ and free Ca^{2+} and Mg^{2+} ions in the slurry. The creation of hydrated $Ca(H_2PO_4)_2$ has been demonstrated to substantially reduce phosphate rock surface hydrophobicity, whereas soluble $Mg(H_2PO_4)_2$ produces a modest suppressing effect on calcite [53,59–61]. Figure 3 illustrates how dissolved metal ions influence mineral surface wettability. Furthermore, multiple investigations have documented that interactions between SO_4^{2-} and Ca^{2+} in slurry result in $CaSO_4(s)$ precipitation, which subsequently adheres to fluorapatite surfaces and hinders NaOL adsorption [62–64].

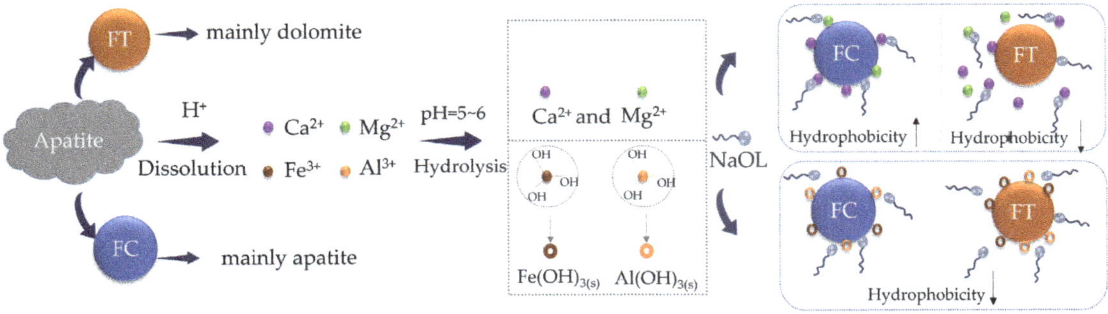

Figure 3. The effect of dissolved metal ions on the surface wettability of minerals [58].

2.2.2. Surface Electrical Properties

The zeta potential serves as an indicator of the electrical double layer and colloidal behavior of nanoparticles, offering insights into their stability, duration in circulation, interactions with proteins, membrane penetration capacity, and biological compatibility. This analytical technique is routinely employed to evaluate the electrical properties at mineral particle surfaces [65]. The positioning ions present on apatite surfaces are Ca^{2+}, $CaOH^+$, HPO_4^{2-}, $H_2PO_4^-$, PO_4^{3-}, H^+, and OH^-. Given the variable geological environments and mineral compositions of apatite from different sources, as well as the mutual substitution of various cations and anions, the point of zero charge (PZC) of apatite exhibits significant variation, from pH = 1 to 8.7 [40,66–73]. The isoelectric point (IEP) is defined as the negative logarithm of the concentration of location ions when the zeta potential is zero [74]. The electrical properties at a mineral's surface or in mineral–reagent complexes depend on multiple factors: the inherent mineral makeup (including mineral variety and impurity content) along with the particular reagents employed (like depressants and collectors), with consideration to both their molecular configuration and concentration levels [75]. In the region to the left of the IEP, the mineral surface carries a positive charge and can be captured by anionic collectors. Conversely, in the right section of the IEP, the mineral surface acquires a negative charge and becomes amenable to collection by positively charged reagents. When the mineral surface and collector molecules share identical charge characteristics, the polar functional groups of the reagent undergo chemical interaction with active sites on the mineral surface, resulting in their attachment to the mineral [51].

Flotation reagents can modify the electrical properties of mineral surfaces (Table 2). For instance, andiroba oil, at a pH of 7.5 and a collector concentration of 20 mg/L, has been demonstrated to induce a negative shift in fluorapatite's IEP curve throughout the complete examined pH spectrum [76]. According to Table 2, each of these collectors caused a reduction in the IEP value and increased negative zeta potential, demonstrating significant reactivity with the target mineral surfaces [77]. In addition, It is evident from the extant literature that metal ions in solution have varying degrees of impact on the surface potential of minerals [78,79]. Liao's study examined how typical chloride-based electrolyte solutions containing Na^+, Mg^{2+}, and Ca^{2+} cations influence the zeta potential of finely ground apatite particles. The study revealed that the surface potential of apatite in the electrolyte solution was significantly higher than that of apatite in deionized water, with divalent Ca^{2+} and Mg^{2+} providing a more positive potential than the monovalent Na^+ [80]. Jiang et al. conducted a study on the effects of Al^{3+} and Fe^{3+} on the surface electrical properties of fluorapatite. The study revealed that the additions of Al^{3+} and Fe^{3+} substantially enhanced the surface charge of apatite across the pH spectrum from 3 to 11 [58].

Table 2. The IEP of apatite with and without the addition of collectors.

Reagents	Concentration	Presence	Absence	Ref.	Year
HOl	7.5 mg/L	<3.0	4.0	[81]	1990
Oleic acid amide	15 mg/L	<2	2.8	[82]	2017
Pataua oil	Unknown	4	6.5	[83]	2019
LH-01	1×10^{-2} mol/L	4.8	4.5	[84]	2024
Frying oil	30 mg/L	<1	3.70	[85]	2024
SDBS	1.33×10^{-1} mol/L	<3	3.08	[86]	2021
Oleamide	2.67×10^{-4} mol/L	<3	3.08	[86]	2021

3. Flotation Reagents

Researchers utilized density functional theory (DFT) to probe the adsorption properties of a single water molecule at metal sites on the surface of apatite. The apatite surface contains unsaturated Ca metal active sites, where water molecules can strongly adsorb. The oxygen and hydrogen atoms of the water molecule form hydroxylated metal active sites and hydrogen bonds of varying strengths with the metal atoms and oxygen on the mineral surface. The surface reconstruction caused by the hydroxylation and hydrogen bonding of water molecules on the mineral surface is the essence of apatite exhibiting hydrophilic properties in natural systems [87]. Typically, negatively charged surfactants are applied in positive flotation to separate phosphate minerals from calcium-containing waste minerals, whereas positively charged surfactants are utilized in negative flotation to eliminate silica-based impurities from phosphate deposits. During positive flotation operations, the pulp's alkalinity is maintained at around pH 9.5 through soda ash addition, while controlled doses of sodium silicate serve to inhibit the floating of silica minerals. Subsequently, the process involves the utilization of anionic surfactants to facilitate the flotation of phosphate minerals. Conversely, in the reverse flotation process, cationic surfactants are utilized to facilitate the flotation of silicates at a weakly acidic or neutral pH, thereby enabling the recovery of phosphate minerals from the tailings product [88].

Apatite, dolomite, and calcite are all calcium-containing minerals. When these minerals dissociate, the surface exposes Ca active sites, which are typically the adsorption sites for flotation reagents. In natural systems, dolomite and calcite exhibit hydrophilic properties similar to those of apatite. Therefore, the differentiation of apatite from calcium-containing minerals like dolomite and calcite is frequently accomplished through an inverse flotation approach, employing fatty acid collectors and suitable apatite depressants. Alternatively, a direct flotation process can be used, employing depressants that selectively attach to dolomite surfaces [89,90]. In the flotation system of apatite and quartz, cationic collectors are commonly utilized. Apatite has higher O and Ca activity, and the Ca atoms on the surface carry a relatively large positive charge. The Ca 3d states in the valence band exhibit strong metallic activity, making it difficult for apatite to interact with cationic collectors [91]. In comparison with apatite, the quartz surface demonstrates a higher absolute value of negative zeta potential. Therefore, cationic collectors exhibit stronger electrical attraction to quartz surfaces, thereby enhancing their interaction with quartz particles in the slurry relative to their behavior with apatite. This improves the efficiency of inverse flotation for separating apatite from quartz [48]. The present review systematically categorizes both collectors and depressants employed in the beneficiation of apatite from its principal associated minerals—dolomite, calcite, and quartz. A particular focus is placed on recently developed reagents. Furthermore, the mechanisms of action of the diverse reagents are analyzed.

3.1. Collector

Collectors, known as "surfactants", are amphiphilic molecules that serve as crucial components in mineral separation by flotation. The "head" of the surfactant is capable of binding to the mineral surface, while the "tail" is hydrophobic and facilitates binding with bubbles, thus enabling the surfactant to rise to the surface [92]. Collectors, functioning as surface-active agents, are classified based on their polar group charge characteristics: anionic (negatively charged), cationic (positively charged), amphoteric (charge varies with pH), and nonionic (electrically neutral) [93]. Figure 4 shows the effect of pH on the dissociation of amphoteric surfactants. At pH values ranging from 6 to 7, the collector manifests amphoteric properties. Consequently, when the pH exceeds 7, the amphoteric collector predominantly behaves as an anion. Numerous chemical reagents serve distinct roles in phosphate ore flotation systems, acting as collectors, depressants, and modifying agents for both positive and reverse flotation operations. During direct flotation, phosphate components are recovered under basic pH environments through the application of negatively charged collectors alongside gangue-depressing chemicals. In contrast, reverse flotation employing cationic amine collectors at near-neutral pH values selectively removes silicate impurities from phosphate matrices. Conversely, anionic reverse flotation describes the use of anionic collectors to float carbonate minerals under acidic pH conditions, while appropriate depressants are employed to suppress phosphate minerals [24]. This section summarizes the different types of apatite collectors and their action mechanisms, primarily utilizing illustrative examples of newly emerging collectors. Table 3 provides a concise list of apatite flotation collectors, with a particular focus on those that have emerged in recent years.

Table 3. Collectors used in apatite ore flotation.

Ionic Property	Collector	pH *	Concentration	Ref.	Year
Anionic	FrOC	9	1000 g/t	[85]	2024
	2-Cl-9-ODA	9	3×10^{-4} mol/L	[94]	2024
	SNLS	10	5×10^{-6} mol/L	[95]	2023
	CSFA	7	40 mg/L	[96]	2021
	BHA	9	1.2×10^{-4} mol/L	[97]	2019
Cationic	DHDB	4.5	15 mg/L	[98]	2019
	LPDC	6.5	25 mg/L	[99]	2023
	HDMEA	6.43	20 mg/L	[54]	2022
Amphoteric	DDALA	6	3×10^{-4} mol/L	[48]	2024
	Lecithin	8.5	40 mg/L	[100]	2023
	C12Giy	5	30 mg/L	[101]	2022
Nonionic	NI-EP	4.5~5.2	125 g/t	[102]	2025
	oxyethylenated cetyl ether	10	360 g/t	[103]	2022

*: The initial pH in slurry, measured before flotation.

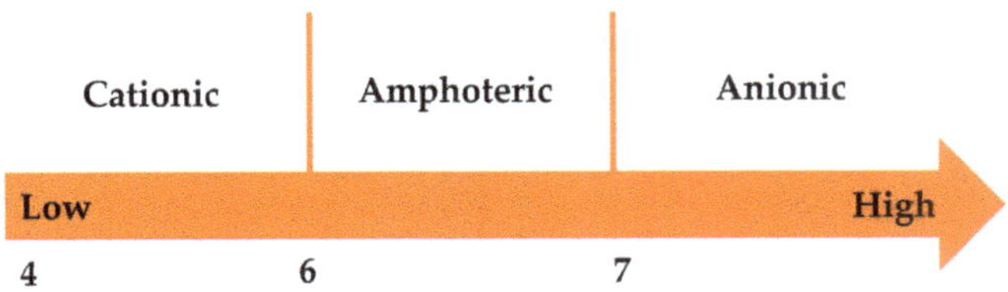

Figure 4. Effect of pH on the dissociation of amphoteric surfactants [104].

3.1.1. Anionic Collector

Fatty acid-based collectors have shown effectiveness in increasing the hydrophobicity of both phosphate minerals and calcium-bearing gangue minerals, including dolomite and calcite. Sodium oleate (NaOL) and oleic acid (HOl), serving as characteristic fatty acid salts and fatty acids, respectively, represent the most commonly employed negatively charged collectors for calcium-bearing minerals, including phosphate minerals [66,105,106]. Fatty acids are classified as carboxylic acid-type collectors, with a general chemical structure formula of R-COOH (Na^+ or K^+). In the context of density functional theory, the electronic structure of oleic acid, a prevalent fatty acid collector, was examined. It was demonstrated that the oxygen atoms present in both the oleic acid molecule and the oleate ion exhibit high levels of reactivity. The oxygen atom is identified as the bonding atom in the interaction between oleic acid and the oxide mineral surface [107]. It has been established that the ionization of a substantial quantity of $RCOO^-$ and $H(RCOO)_2^-$ results in the occurrence of chemical adsorption at the Ca active sites on the mineral surface. This, in turn, leads to a modification in the surface's hydrophobicity [108]. However, because $RCOO^-$ and $H(RCOO)_2^-$ ions can indiscriminately undergo non-specific adsorption with metal active sites such as Ca^{2+}, Fe^{3+}, and Mg^{2+}, and due to the simple geometric configuration of the carboxyl group, which cannot differentiate the coordination environment of cations on mineral surfaces through steric hindrance effects, these reagents lack selectivity [109,110]. In addition, a high dosage of collectors is required, and factors such as low temperature, the presence of cations, and slimes have a significant impact on their performance. Moreover, their performance tends to fluctuate, as their composition—being derived from natural sources—may vary with the seasons [26,111,112]. The chemical alteration of fatty acids (including sulfidation, sulfonation, and esterification processes) and combining diverse fatty acid configurations with negatively charged, zwitterionic, and uncharged collectors are regarded as viable strategies [24,113,114]. By modifying fatty acids to incorporate a greater variety of functional groups, their selectivity and solubility can be improved. This modification can also enhance the dispersion state of fatty acids and promote their adsorption onto mineral surfaces [24,115]. The combined use of reagents with diverse molecular structures can diversify adsorption mechanisms and simultaneously improve frothing characteristics. Figure 5 illustrates the process by which mixed reagents strengthen collector attachment to mineral surfaces [116,117].

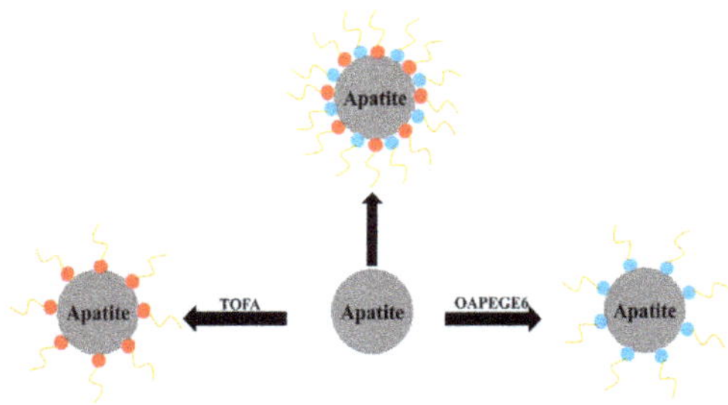

Figure 5. Adsorption of mixed reagents on the surface of apatite. TOFA: Tall oil fatty acid; OAPEGE6: Oleic acid polyethylene glycol ester [32].

3.1.2. Cationic Collectors

Cationic collectors are mainly employed in the reverse flotation separation of silica-based minerals, but they can also be used to float dolomite and calcareous minerals from phosphate ores [24]. Amine-based cationic collectors, including dodecylamine (DDA) and its derivatives such as dodecylamine hydrochloride, are conventional cationic collectors that have been extensively utilized for the treatment of silicate minerals [118,119]. There are three distinct adsorption mechanisms for DDA on the quartz surface. First, monolayer adsorption occurs at low concentrations. Second, surface micelle formation is observed at moderate concentrations. Third, multilayer adsorption is evident at high concentrations, accompanied by the precipitation of molecular amines [120]. The DDA's hydrocarbon chain adopts a near-vertical orientation relative to the apatite surface, enabling the effective dispersion of its non-polar segment in the flotation pulp. This configuration facilitates DDA accumulation on apatite through combined hydrophobic effects and intermolecular forces [121]. Figure 6 illustrates the proposed adsorption configuration of DDA on apatite. However, challenges including excessive froth formation, difficult foam control, and DDA's poor mineral specificity often result in unsatisfactory separation outcomes [122–124]. These limitations have stimulated growing research interest in advanced amine-type collector development. Studies have shown that, when ester and ether groups are inserted into amines, the interaction strength increases. In addition, the binding energy generally rises with longer alkyl chains but stabilizes beyond a specific carbon count [125]. The isopropyl group separating the hydroxyl and secondary amine in DDAIP creates spatial constraints that inhibit concurrent coordination of the molecule's oxygen and nitrogen atoms with surface Ca^{2+} sites on phosphate minerals. Consequently, DDAIP demonstrates superior target mineral specificity compared to DDA when separating phosphate from quartz through reverse flotation processes [126].

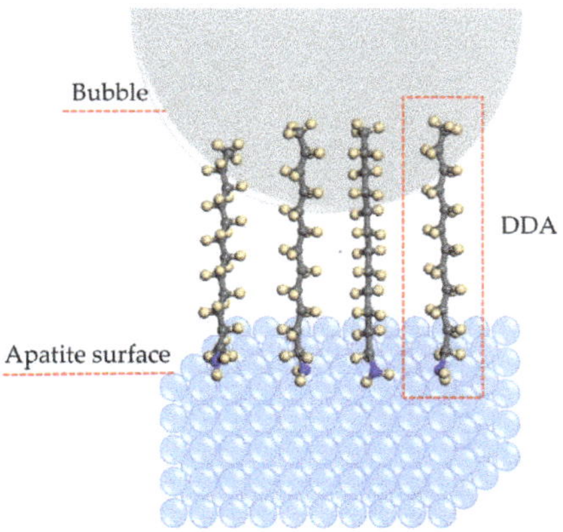

Figure 6. Schematic of the adsorption model of DDA on the apatite surface [121].

3.1.3. Amphoteric Collectors

In previous flotation experiments, researchers found that amphoteric surfactants exhibited greater selectivity than sodium oleate and were less sensitive to dissolved ions [127]. Most amphoteric surfactants in solution exist in three distinct forms (cationic, amphoteric, and anionic) when undergoing dissociation, and their relative concentrations change with variations in the solution pH [100]. Indeed, amphoteric surfactants predominantly behave as cations at low pH values [104]. Therefore, under acidic conditions (pH 4–5), zwitterionic molecules become the dominant species in solution and attach to quartz's negatively charged surfaces through their protonated amine groups. Consequently, at a pH smaller than 6, the amphoteric ions can act as cations, thereby reversing the surface charge of quartz and electrostatically adsorbing. In contrast, minerals such as calcite, which possess a positive surface charge, experience greater repulsion and consequently adsorb less [101].

Consequently, quartz flotation is reduced, while carbonate flotation is increased. Lecithin, an environmentally friendly amphoteric foaming agent, functions as a collecting agent in the reverse flotation beneficiation of phosphate ores [100]. Under conditions of pH 8.5 and 40 mg/L lecithin dosage, successful differentiation between phosphate-bearing apatite and carbonate impurities (calcite and dolomite) can be attained, resulting in a P_2O_5 grade of 32.39% with 96% mineral recovery [100]. Figure 7 shows the separation mechanism of phosphate rock using lecithin-based reagents. The newly developed amphoteric collector, N-dodecyl-β-alanine (DDALA), structurally modified from dodecylamine through the introduction of a propionic acid functional group, predominantly binds to quartz surfaces via its positively charged $DDALA^+$ species. For apatite and dolomite surfaces, the neutral DDALA molecular form serves as the primary active agent. Flotation experiments reveal that DDALA, when combined with a xanthogenate (XG) depressant, successfully enhances the inverse flotation differentiation between apatite and quartz. Additionally, in conjunction with oleate collectors, DDALA shows exceptional performance in differentiating apatite from dolomite via reverse flotation techniques [48].

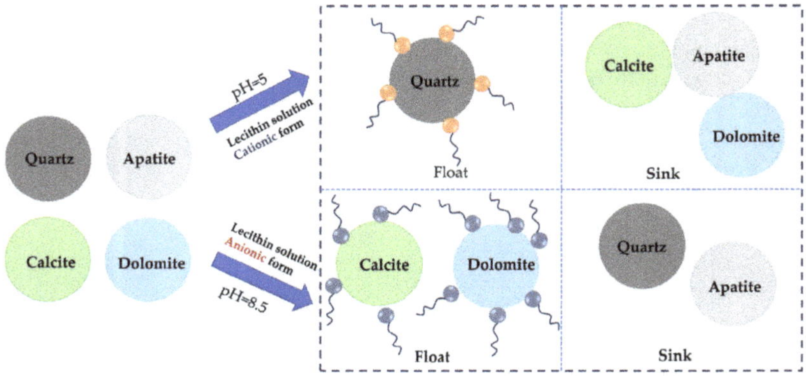

Figure 7. Separation mechanism of phosphate ore using a lecithin solution [100].

3.1.4. Nonionic Collector

Nonionic collectors may consist of unmodified aliphatic alcohols or their chemically altered variants, such as ethoxylated alcohols [24]. Nonionic collectors are typically combined with other ionic collectors to enhance flotation performance. The prevailing opinion among researchers is that the improved flotation performance achieved using these mixed collector systems stems from the simultaneous adsorption of oppositely charged or neutral collectors on mineral surfaces. This phenomenon decreases charge-based repulsion among adsorbed collectors while enhancing both collector adsorption density and intermolecular hydrophobic interactions between adjacent hydrocarbon tails. This leads to a denser packing of the collectors [128–130]. Studies have found that, after adding fatty alcohols, the fatty alcohols co-adsorb with oleic acid on the mineral surface, with their quantity significantly higher than when fatty alcohols are used alone, while the adsorption of oleic acid remains similar. Consequently, the recovery of fluorapatite and calcite during oleic acid flotation is significantly improved after the co-adsorption of fatty alcohols [102,131]. Simultaneously, the amalgamation of amine collectors and fatty alcohols has shown effectiveness in improving the flotation performance of quartz [102]. In addition, ether and ester nonionic surfactants, such as polyethylene glycol alkyl ethers and Tween 80, can also improve the effectiveness of fatty acid collectors in the apatite flotation system [103,132].

3.2. *Depressant*

As previously stated, apatite and its gangue minerals (e.g., dolomite, calcite, and quartz) are all hydrophilic. Furthermore, the calcium-active sites shared by apatite with dolomite and calcite, in addition to the calcium ions dissolved from apatite being able to adhere to the surface of quartz minerals, serve to reduce the selectivity of collectors. Consequently, the utilization of depressants within the apatite flotation system becomes paramount [133]. The basic principle for selecting depressants is to modify the hydrophobic groups of the collector to hydrophilic groups based on the specific mineral being targeted. Depending on the flotation process and gangue type, the main types of depressants include carbonate depressants, silicate depressants, and phosphate depressants (Table 4) [24,134].

Table 4. Depressants used in the flotation of apatite.

Mineral to Be Depressed	Depressant	Concentration	pH *	Ref.	Year
Apatite	H_3PO_4	80 mg/L	7.6	[135]	2017
	Sodium sulphate	3000 g/t	6.0	[136]	2014
	Phosphorylated starch	800 g/t	9.0	[137]	2025
	Tragacanth gum	30 mg/L	8.0	[138]	2025
Carbonate minerals	Al starch	10 mg/L	9.0	[27]	2025
	Pectin	80 mg/L	9.0	[139]	2023
	Gum Arabic	75 mg/L	9.0	[140]	2022
	Carboxymethyl chitosan	10 mg/L	9.0	[141]	2020
	Konjac glucomannan	40 mg/L	9.0	[142]	2024
Silicate minerals	Citric acid	1×10^{-3} mol/L	12	[143]	2019
	Sodium silicate	500 g/t	9.4	[144]	1990

*: The initial pH in slurry, measured before flotation.

3.2.1. Depressors of Apatite

Phosphate depressants mainly function in apatite reverse flotation to suppress apatite. Commonly used inhibitors include sulfuric acid, phosphoric acid, and their related compounds [55]. The depressant mechanism is attributed to the ionization of sulfuric acid and phosphoric acid into $H_2PO_4^-$ and HSO_4^- in the acidic slurry environment, which bind to the Ca^{2+} exposed on the apatite surface, creating a hydrophilic coating that increases the mineral's water affinity. This simultaneously reduces the number of accessible reactive centers on the apatite surface, consequently diminishing collector attachment [135,145–150]. From the above, it can be seen that the formation of $Ca(H_2PO_4)_2$ and $Ca(HSO_4)_2$ enhances apatite hydrophilicity, leading to its depression. It can be concluded that sulphate and phosphate can also act as depressants for apatite [136,151,152]. However, it is worth noting that, although sulfuric acid and its derivatives have a strong depressant effect on phosphate minerals and their advantage is their relatively low cost, they can cause pipeline blockage due to gypsum formation [134].

With the increasing awareness of environmental protection, researchers have begun to focus on the development of environmentally friendly depressants, such as organic acids, polymers, and their derivatives, including nitrotrimethylphosphonic acid, modified starch, and polyacrylamide [153–155]. Dai et al. studied a novel depressant, TG, which is a polysaccharide. They investigated its depressant effect on apatite and its mechanism in a sodium oleate (NaOL) system. The enhanced depressing action of TG on apatite results from more robust and durable chemical bonds established between the –COOH group of TG and the Ca^{2+}-binding sites on the apatite surface. This potent TG–apatite association effectively blocks subsequent NaOL adsorption [138]. In their investigation, Oulkhir et al. examined the depressant effect of phosphorylated starch (PS) extracted from potato waste, a material that has been posited as a green and effective depressant. PS exhibits significant surface attachment to apatite via chemical bonding between its phosphate functional groups and reactive calcium centers on the mineral surface [137].

3.2.2. Depressors of Carbonate Minerals

Calcite and dolomite are the primary carbonate gangue minerals present in apatite ores. Common carbonate mineral depressants include lignosulfonates, sulfonated phenolic resins, nitrated humic acid salts, sulfonated naphthalene (S711), and sulfonated anthracene (S808). Recent studies have increasingly concentrated on the development of new silicate depressants that exhibit high selectivity, environmental friendliness, and cost-effectiveness. Polysaccharides, recognized for their environmental sustainability, have become a significant focus of investigation within the domain of bio-based depressants. The presence of -OH and -COO groups in polysaccharides has been demonstrated to enable complexation with metallic cations (particularly Ca^{2+} and Mg^{2+}) present on mineral surfaces in phosphate deposits, including apatite, calcite, and dolomite [134,156]. Examples include sodium alginate, carboxymethyl cellulose, pectin, gum Arabic (GA), xanthan gum (XG), and carboxymethyl chitosan derivatives [139–141,157,158]. Figure 8 shows several molecular structural formulas of typical bio-based depressants. Figure 9 presents a schematic representation of the molecular interactions occurring between polysaccharide compounds and the surface structures of calcite, dolomite, and apatite minerals.

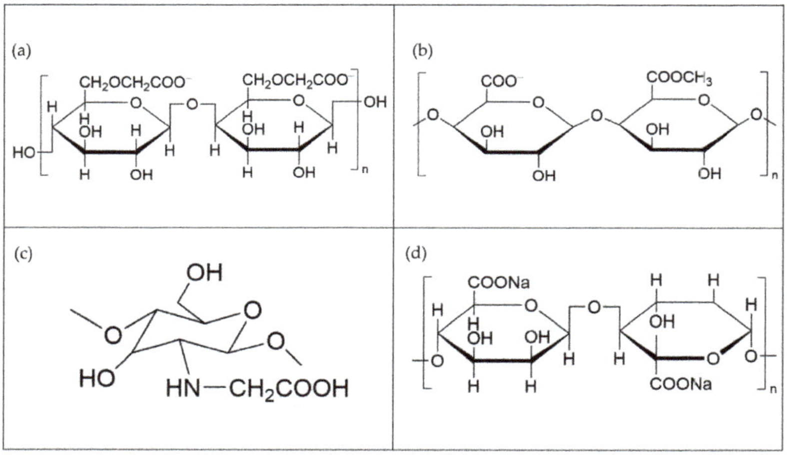

Figure 8. Molecular structural formulas of typical bio-based depressors [95]. (**a**) Carboxymethyl cellulose; (**b**) pectin; (**c**) carboxymethyl chitosan; and (**d**) sodium alginate.

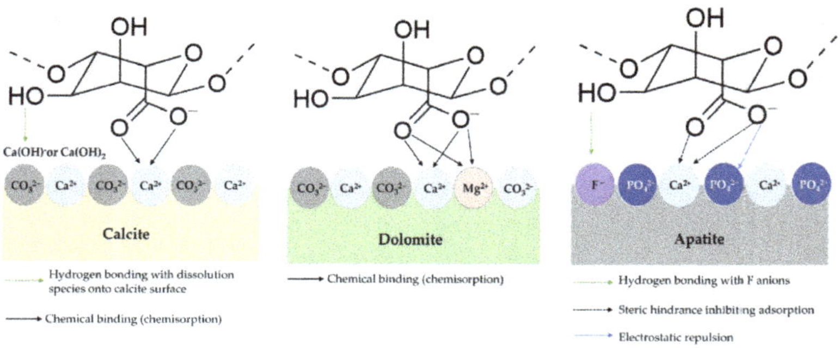

Figure 9. The schematic illustration of molecular interactions occurring between polysaccharide compounds and the surfaces of calcite, dolomite, and apatite minerals [156].

The surfaces of bacteria are covered with various structures, including peptidoglycan, teichoic and teichuronic acids, lipoproteins and lipopolysaccharides, surface protein arrays, and bacterial capsule polymers, which are typically polysaccharides or polypeptides. These structures are capable of binding with metal ions [159]. The use of microorganisms as depressants for carbonate minerals in apatite flotation has gradually become a research hotspot. Bacteria such as *Bacillus subtilis, Mycobacterium, Rhodococcus erythropolis CD 130, Pseudomonas fluorescens, Escherichia coli, Bacillus subtilis, C. albicans,* and *Escherichia coli* have all been shown to preferentially adsorb on the dolomite's surface, competing with the collectors for adsorption on dolomite to achieve selective suppression in the apatite flotation process [145,160,161].

3.2.3. Depressors of Silicate Minerals

Sodium silicate represents the most widely employed depressing agent for siliceous apatite, used to suppress silicate minerals such as quartz. However, Wang et al. demonstrated that, in a fatty acid flotation system utilizing sodium silicate depression, achieving high-purity apatite concentrates proves challenging [162]. The primary rationale for this phenomenon pertains to the attachment of calcium ions (Ca^{2+}) that have been dissolved from the apatite mineral to the surface of quartz. This process serves to activate the quartz, thereby diminishing the selectivity of the collector. Hence, the selection of an effective depressant to inhibit Ca^{2+}-activated quartz is imperative. In the context of the apatite flotation system, the mechanism of action of silicate depressants can be primarily divided into two distinct aspects. Firstly, they compete for adsorption on the quartz surface, reducing the adsorption of the collector. Secondly, they interact with Ca^{2+}, reducing its adsorption on the quartz surface, decreasing the activation of the quartz surface, and hindering the interaction between the collector and the quartz surface [162]. Sayilgan's research examined the impact of carbonates on the flotation behavior of quartz under alkaline conditions, utilizing amine collectors and sodium oleate in the presence of calcium ions [163]. The research results show that carbonates form precipitates with Ca^{2+} ions in the solution, significantly reducing the Ca^{2+} concentration and decreasing the activation of quartz. Wang used citric acid (CA) as a depressant for sodium NaOL flotation of calcium-activated quartz and analyzed its depression mechanism [143]. CA's inhibitory effect on activated quartz operates through dual pathways: initially, CA anions remove surface-adsorbed Ca^{2+} ions from quartz, decreasing available attachment sites for NaOL; subsequently, CA molecules bound to residual Ca^{2+} sites block further NaOL adsorption on the quartz surface. In recent years, many depressants for calcium gangue minerals have been studied to explore their depressing effect on quartz, such as sodium alginate, carboxymethyl chitosan, modified starch, pectin, and carboxymethyl cellulose [164–168].

4. Interface Regulation of Fine Apatite

The beneficiation of phosphate ore presents a challenge in the recovery of fine-grained apatite from low-grade and medium-grade collophane, weathered phosphate mud, and phosphate tailings. Studies have shown that particles of different sizes behave differently during flotation, directly affecting both the flotation recovery and flotation rate. For phosphate minerals, the optimal flotation particle size range is 60–200 μm. When the particle size falls below 20 μm, the recovery rate sharply decreases to below 70% [169]. It is imperative to strengthen the research on interface regulation for fine-grained apatite flotation to enhance the flotation process and promote the efficient utilization and sustainable development of medium- and low-grade fine phosphate resources. The inferior flotation performance of fine mineral particles principally stems from their reduced probability of contacting standard-sized flotation bubbles moving at typical velocities [169,170]. This bubble-particle attachment likelihood is principally governed by the comparative sizes of the gaseous and solid phases, illustrated in Figure 10.

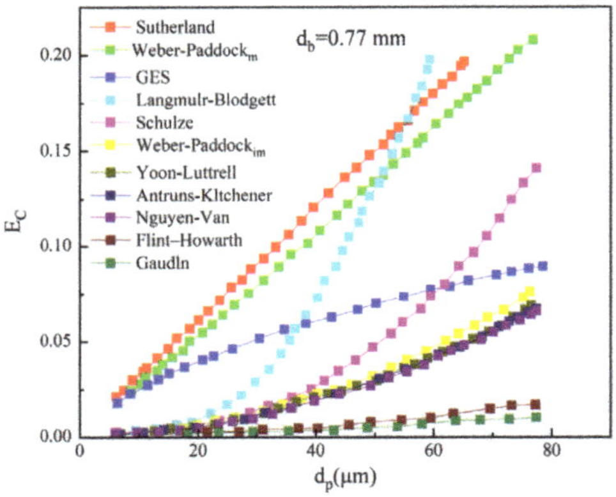

Figure 10. Collision efficiencies as a function of particle size. E_c—Collision efficiency; d_p—particle diameter; d_b—bubble diameter [171,172].

The probability of successful attachment rises proportionally with growing particle dimensions. Furthermore, elevated agitation rates provide greater energy input, accelerating particle movement and consequently improving particle–bubble contact frequency. And the strong energy input creates strong turbulence in the slurry, and the enhancement in turbulence further reduces the size of bubbles and also increases the probability of collision of fine particles [173]. Interface regulation during flotation is crucial, including the regulation of the mineral/water interface wettability by flotation reagents (collectors and modifiers), the regulation of interactions between mineral particles (dispersion of heterogeneous particles and hydrophobic aggregation of homogeneous particles), and the regulation of interactions between mineral particles and bubbles (collision, adhesion, and detachment) [174]. The flotation reagents have been introduced earlier, and this paper will continue to summarize recent research on the interactions between mineral particles and the interactions between mineral particles and bubbles.

4.1. Interaction Regulation Interparticle: Particle Aggregation and Stability

The targeted agglomeration of fine particles to enlarge their effective size represents a viable approach for enhancing fine mineral recovery during flotation [175]. Understanding interparticle forces is fundamental for controlling the selective aggregation processes of finely divided minerals. Ye [174] conducted a study on the dispersion and hydrophobic agglomeration regulation of fine-grained apatite particles in an apatite–quartz system under various dispersants (sodium silicate, sodium hexametaphosphate, sodium carbonate) and emulsified kerosene. The study established that, under natural conditions, fine apatite exhibited inadequate dispersion stability, while fine quartz demonstrated robust dispersion stability. In the presence of an alkaline environment, the dispersion stability of fine-grained apatite exhibited a modest enhancement. The utilization of dispersants led to a substantial augmentation in the dispersion stability of fine apatite, thereby impeding the process of heterogeneous aggregation. Their dispersion ability was ranked from highest to lowest as follows: sodium hexametaphosphate > sodium silicate > sodium carbonate. Furthermore, the micron-sized kerosene droplets prepared by mechanical stirring acted as bridging agents between the hydrophobic apatite particles induced by sodium oleate, thereby increasing the size and strength of the agglomerates and subsequently enhancing the flotation rate of fine-grained apatite. Research has identified an interdependent relationship among particle agglomeration, slurry viscosity, and interparticle forces, analogous to a "three-factor equilibrium". Elevated slurry viscosity reflects strengthened particle interactions, predominantly resulting from intensified hydrophobic attraction. This subsequently promotes more extensive clustering of apatite particles. Moreover, as apatite aggregation progresses, a three-dimensional network develops within the pulp, impeding liquid movement and consequently raising the system's apparent viscosity [176].

4.2. Interaction Regulation of Particle–Bubble

Flotation relies on the selective adhesion of specific mineral particles to gas bubbles, making bubble–particle attachment a critical stage in the separation process, determining the efficiency and selectivity of the process [177,178]. The process of particle capture by bubbles in the slurry is characterized by a series of physical interactions, including collision, attachment, and detachment [179,180]. Angelica investigated the influence of bubble size distribution on fine apatite particles in flotation columns. The research involved multiple flotation experiments designed to evaluate different bubble size distribution profiles. The test series included small bubbles ($d_{32} < 300$ μm), medium-sized bubbles ($300 < d_{32} < 1000$ μm), and large bubbles ($d_{32} > 1000$ μm). The optimal bubble size range for fine apatite flotation was determined. When using small bubbles ($d_{32} < 300$ μm) to float fine apatite, the highest P_2O_5 content could be obtained, but the P_2O_5 recovery rate was very low (40%) under these conditions. In contrast, using larger bubbles ($d_{32} > 1000$ μm) resulted in a lower P_2O_5 content, but the flotation achieved the highest P_2O_5 recovery. Considering both factors (P_2O_5 content and recovery rate), the best flotation performance was achieved using medium-sized bubbles. Various methods have been examined to enhance the collision effectiveness between bubbles and fine particles, such as modifying aeration system operating parameters, enhanced flocculation flotation techniques, pulsed aeration flotation processes, dissolved air flotation (DAF) systems, electrostatic interaction-based flotation with colloidal gas bubbles, and nanobubble-supplemented flotation approaches [179,181–185]. Santana et al. [182] investigated fine apatite particle separation using DAF technology. Their comparative analysis revealed DAF's superior performance over traditional flotation methods, demonstrating extended bubble retention periods, substantially greater bubble quantities per unit air volume, and expanded bubble surface areas, which all contribut to improved particle–bubble interaction probabilities. Jiang et al. [186] investigated how impeller rotation rates influence nanobubble flotation performance. Their findings revealed that higher impeller speeds produced larger aggregate sizes, with rotational velocities between 1200 and 2800 rpm generating substantially greater quantities and sizes of nanobubble clusters compared to conventional flotation systems, which was consistent with the improvements in flotation recovery and flotation rate. Additionally, Zeng reported that nanobubbles could promote the recovery of apatite using sodium oleate, but had the opposite effect when using dodecylamine [187]. Researchers have utilized the cavitation effect and acoustic radiation force effect of ultrasound to cause bubble rupture, which impacts particles and bubbles, thereby achieving the movement, dispersion, and coalescence of particles and bubbles [188,189]. Jung et al. [190] developed a microbubble generation device and found that, within an acoustic field, adjacent microbubbles tend to cluster together, forming bubble aggregates that enhance the buoyancy of the bubbles. Gao et al. examined how varying gas supply parameters influence bubble characteristics in ultrasonic-assisted flotation systems. Their study showed that, by altering the inner diameter and angle of the gas generator, the initial morphology and motion paths of the bubbles could be modified, ultimately influencing their flotation performance [191].

5. Future Outlook

For future research on apatite flotation interface control, several considerations can be made for researchers:

(1) Strengthening the development of highly selective, environmentally friendly reagents and expanding research from micro-flotation experiments to industrial trials.
(2) Mixed reagents have been proven to perform better than single reagents, but their mechanisms are not yet thoroughly understood. In-depth research should be conducted using advanced detection equipment, and based on this, more reagent combinations should be proposed.
(3) Research on the relationship between reagent structure and particle size in flotation systems is still insufficient, especially for fine particles or even ultra-fine particles (less than 10 μm).
(4) Research on particle–particle interactions is still at the pure mineral testing stage, while the actual flotation slurry environment is much more complex. There should be more focus on the study of particle interactions in complex environments.
(5) Regarding bubble size regulation, the optimization and development of micro-bubble generation equipment should be strengthened.

It should be emphasized that all research is based on the properties of mineral surfaces, and more importantly, the core of research work should focus on further studies into mineral surface properties to enhance the comprehension of reagent adsorption mechanisms and the intricate relationships between solid particles and gas bubbles.

6. Conclusions

(1) Apatite is widely used in various industrial fields, and its demand is continuously increasing. Flotation is the most commonly used process for enriching apatite. The main challenges in the flotation enrichment process of apatite are twofold. First, developing highly selective reagents to enhance the surface wettability differences between apatite and gangue minerals such as calcite, dolomite, and quartz. Second, regulating the interactions between particles and between particles and bubbles to increase the probability of particle–bubble collisions.

(2) The flotation interface regulation reagents for apatite mainly include collectors and depressants. The types of collectors primarily include anionic, cationic, amphoteric, and nonionic collectors. Developing highly selective collectors is crucial. The chemical alteration of fatty acids or combining diverse fatty acid configurations with manufactured negatively charged, zwitterionic, and uncharged collectors represents a primary strategy. Cationic collectors are widely used in silicate flotation. Studies of positively charged collectors primarily examine how varying hydrocarbon chain lengths and polar groups influence collection efficiency. Uncharged collectors are typically employed alongside ionic collectors to enable cooperative adsorption, which strengthens mineral surface hydrophobicity and enhances separation effectiveness.

(3) According to flotation processes and gangue types, the main types of depressants include carbonate, silicate, and phosphate depressants. Commonly used depressants include sulfuric acid, hydrochloric acid, lignosulfonates, and sodium silicate. With the growing awareness of green and sustainable development, organic and bio-based depressants are being continuously developed, such as modified starch, pectin, sodium alginate, and carboxymethyl chitosan. These depressants achieve their inhibitory effects through attachment to calcium-reactive surface sites via numerous -OH and -COO groups.

(4) Two primary strategies for improving fine apatite–bubble collision rates involve modifying particle–particle and particle–bubble interactions to either enlarge particle dimensions or reduce bubble size. Flotation reagents can intensify hydrophobic attraction among mineral particles, promoting hydrophobic agglomeration and effectively increasing the mineral particle size. Advanced techniques like aeration system optimization, enhancing flocculation flotation, oscillating air supply flotation, dissolved air flotation, and the

use of nanobubbles can be employed to enhance the particle–bubble interaction frequency and consequently boost separation performance.

(5) Subsequent investigations should concentrate on the development of highly selective and environmentally friendly reagents, based on a thorough investigation of reagent structures and their mechanisms of action. In addition, more emphasis should be placed on the application of emerging interface regulation technologies in industrial operations, serving as a foundation for the development and optimization of interface regulation devices.

Author Contributions: Conceptualization, writing—review and editing, Z.L. (Zhe Liu); Funding acquiring, conceptualization, writing—review and editing, supervision L.L.; validation Z.L. (Zhuguo Li); formal analysis, M.W.; investigation H.M.; visualization, F.L. All authors have read and agreed to the published version of the manuscript.

Funding: This research was funded by the National Natural Science Foundation of China, grant number 52174243, and Key Research and Development Project of Liaoning Province, grant number 2024JH2/102400026.

Data Availability Statement: All data included in this study are available by contacting the corresponding author.

Conflicts of Interest: Authors Zhuguo Li and Meng Wang were employed by the company Chengde Jingcheng Mining Group Co., Ltd. The remaining authors declare that the research was conducted in the absence of any commercial or financial relationships that could be construed as a potential conflict of interest.

Abbreviations

The following abbreviations are used in this manuscript:

MAP	Monoammonium phosphate
DAP	Diammonium phosphate
(IMA-CNMNC)	The Commission on New Minerals, Nomenclature and Classification of the International Mineralogical Association
AMCSD	American Mineralogist Crystal Structure Database
WCR	Washburn capillary rise
DDA	Dodecylamine
DDAIP	1-(dodecylamine)-2-propanol
PZC	Point of zero charge
IEP	Isoelectric point
DFT	Density functional theory
SDS	Sodium dodecyl sulfate
TOFA	Tall oil fatty acid
OAPEGE	Oleic acid polyethylene glycol ester
HDMEA	N-(2-hydroxy-1, 1-dimethylethyl)
DTAB	Dodecyltrimethyl ammonium bromide
DDALA	N-dodecyl-β-alanine
XG	Xanthogenate
TG	Tragacanth gum
PS	Phosphorylated starch
CA	Citric acid
S711	Sulfonated naphthalene
S808	Sulfonated anthracene
GA	Arabic gum
KGM	Konjac glucomannan

References

1. Mineral Commodity Summaries. *Mineral Commodity Summaries 2024*; Mineral Commodity Summaries: Reston, VA, USA, 2024; p. 212.
2. Grohol, M.; Veeh, C. Study on the Critical Raw Materials for the EU 2023—Final Report. 2023. Available online: https://op.europa.eu/en/publication-detail/-/publication/57318397-fdd4-11ed-a05c-01aa75ed71a1 (accessed on 31 March 2025).
3. Fournie, T.; Rashwan, T.L.; Switzer, C.; Gerhard, J.I. Phosphorus Recovery and Reuse Potential from Smouldered Sewage Sludge Ash. *Waste Manag.* **2022**, *137*, 241–252. [CrossRef] [PubMed]
4. Wei, Z.; Zhang, Q.; Wang, X. New Insights on Depressive Mechanism of Citric Acid in the Selective Flotation of Dolomite from Apatite. *Colloids Surf. A Physicochem. Eng. Asp.* **2022**, *653*, 130075. [CrossRef]
5. Zhang, F.; Williamson, B.J.; Broom-Fendley, S. Apatite Texture and Composition in the Tonglushan Porphyry-Related Skarn System, Eastern China: Implications for Mineral Exploration. *Ore Geol. Rev.* **2023**, *158*, 105493. [CrossRef]
6. Yang, H.; Zhao, Z.; Cao, X.; Fan, H.; Xiao, J.; Xia, Y.; Zeng, M. Geochemistry of Apatite Individuals in Zhijin Phosphorites, South China: Insight into the REY Sources and Diagenetic Enrichment. *Ore Geol. Rev.* **2022**, *150*, 105169. [CrossRef]
7. Dong, L.; Cui, Y.; Qiao, L.; Lan, S.; Zheng, Q.; Shen, P.; Liu, D. A Critical Review on the Flotation of Calcium-Containing Minerals. *Sep. Purif. Technol.* **2025**, *360*, 131082. [CrossRef]
8. Mineral Commodity Summaries. *Mineral Commodity Summaries 2023*; Mineral Commodity Summaries: Reston, VA, USA, 2023; p. 210.
9. Cloud, P. Phosphorite Sedimentation: Phosphate Deposits of the World. *Science* **1987**, *236*, 1125–1126. [CrossRef] [PubMed]
10. Sokolov, A.S.; Frolov, A.A.; Belov, S.V. Regularities in the Distribution and Genetic Features of Phosphate Ore Deposits. *Geol. Ore Depos.* **2001**, *43*, 150–161.
11. Andersson, S.S.; Wagner, T.; Jonsson, E.; Fusswinkel, T.; Whitehouse, M.J. Apatite as a Tracer of the Source, Chemistry and Evolution of Ore-Forming Fluids: The Case of the Olserum-Djupedal REE-Phosphate Mineralisation, SE Sweden. *Geochim. Et Cosmochim. Acta* **2019**, *255*, 163–187. [CrossRef]
12. Zhong, Y.; Liu, G.; Yang, X. Isolating Elemental Phosphorus from Sewage Sludge Ash by Electrochemistry. *Resour. Conserv. Recycl.* **2023**, *190*, 106815. [CrossRef]
13. Marques, C.F.; Olhero, S.; Abrantes, J.C.C.; Marote, A.; Ferreira, S.; Vieira, S.I.; Ferreira, J.M.F. Biocompatibility and Antimicrobial Activity of Biphasic Calcium Phosphate Powders Doped with Metal Ions for Regenerative Medicine. *Ceram. Int.* **2017**, *43*, 15719–15728. [CrossRef]
14. Abouzeid, A.-Z.M.; Negm, A.T.; Elgillani, D.A. Upgrading of Calcareous Phosphate Ores by Flotation: Effect of Ore Characteristics. *Int. J. Miner. Process.* **2009**, *90*, 81–89. [CrossRef]
15. Abouzeid, A.Z.M.; El-Jallad, I.S.; Orphy, M.K. Calcareous Phosphates and Their Calcined Products. *Miner. Sci. Eng.* **1980**, *12*, 73–83.
16. Abouzeid, A.-Z.M. Physical and Thermal Treatment of Phosphate Ores—An Overview. *Int. J. Miner. Process.* **2008**, *85*, 59–84. [CrossRef]
17. Raiymbekov, Y.; Besterekov, U.; Abdurazova, P.; Nazarbek, U. Review of Methods and Technologies for the Enrichment of Low-Grade Phosphorites. *Rev. Inorg. Chem.* **2022**, *42*, 385–395. [CrossRef]
18. Wei, Z.; Sun, W.; Wang, P.; Liu, D.; Han, H. A Novel Metal–Organic Complex Surfactant for High-Efficiency Mineral Flotation. *Chem. Eng. J.* **2021**, *426*, 130853. [CrossRef]
19. Feng, J.; Yang, B.; Zhu, H.; Zhang, H.; Zeng, M. The Utilization of 2-Phosphonobutane-1,2,4-Tricarboxylic Acid (PBTCA) as a Novel Sphalerite Depressant in the Selective Flotation of Galena from Sphalerite. *Appl. Surf. Sci.* **2021**, *569*, 150950. [CrossRef]
20. Jiang, K.; Liu, J.; Wang, Y.; Zhang, D.; Han, Y. Surface Properties and Flotation Inhibition Mechanism of Air Oxidation on Pyrite and Arsenopyrite. *Appl. Surf. Sci.* **2023**, *610*, 155476. [CrossRef]
21. Li, K.; Zhang, H.; Peng, T.; Liu, C.; Yang, S. Influences of Starch Depressant with the Various Molecular Structure on the Interactions between Hematite Particles and Flotation Bubbles. *Colloids Surf. A: Physicochem. Eng. Asp.* **2022**, *652*, 129814. [CrossRef]
22. Xie, H.; Jin, Y.; Zhang, P.; Liu, Y.; Gao, L.; Feng, Q.; Liu, D. Surface Modification Mechanism of Galena with H2SO4 and Its Effect on Flotation Separation Performance. *Appl. Surf. Sci.* **2022**, *579*, 152129. [CrossRef]
23. Gomez-Flores, A.; Heyes, G.W.; Ilyas, S.; Kim, H. Prediction of Grade and Recovery in Flotation from Physicochemical and Operational Aspects Using Machine Learning Models. *Miner. Eng.* **2022**, *183*, 107627. [CrossRef]
24. Sis, H.; Chander, S. Reagents Used in the Flotation of Phosphate Ores: A Critical Review. *Miner. Eng.* **2003**, *16*, 577–585. [CrossRef]
25. Somasundaran, P.; Sivakumar, A. Advances in Understanding Flotation Mechanisms. *Min. Metall. Explor.* **1988**, *5*, 97–103. [CrossRef]
26. Derhy, M.; Taha, Y.; Benzaazoua, M.; El-Bahi, A.; Ait-Khouia, Y.; Hakkou, R. Assessment of the Selective Flotation of Calcite, Apatite and Quartz Using Bio-Based Collectors: Flaxseed, Nigella, and Olive Oils. *Miner. Eng.* **2022**, *182*, 107589. [CrossRef]

27. Wang, R.; Zhang, H.; Xiao, J.; Sun, W.; Tao, D.; Zuo, Y.; Han, H. Novel Metal-Starch Depressant for the Flotation Separation of Ultrafine Apatite and Dolomite Based on Experimental and Calculational Study. *Sep. Purif. Technol.* **2025**, *362*, 131877. [CrossRef]
28. Wang, L.; Zhang, X.; Xue, N.; Li, Z.; Zhu, Y.; Nie, Y. Flotation Separation of Apatite from Dolomite with Tara Gum as Depressant: An Experimental and Molecular Dynamics Simulation Study. *Colloids Surf. A Physicochem. Eng. Asp.* **2025**, *708*, 136044. [CrossRef]
29. Liu, C.; Zhu, L.; Fu, W.; Chi, R.; Li, H.; Yang, S. Investigations of Amino Trimethylene Phosphonic Acid as a Green and Efficient Depressant for the Flotation Separation of Apatite from Calcite. *Miner. Eng.* **2022**, *181*, 107552. [CrossRef]
30. Ruan, Y.; He, D.; Chi, R. Review on Beneficiation Techniques and Reagents Used for Phosphate Ores. *Minerals* **2019**, *9*, 253. [CrossRef]
31. Huang, X.; Zhang, Q. Interaction Behavior between Coarse and Fine Particles in the Reverse Flotation of Fluorapatite and Dolomite. *Langmuir* **2023**, *39*, 12931–12943. [CrossRef]
32. Yang, B.; Cao, S.; Zhu, Z.; Yin, W.; Sheng, Q.; Sun, H.; Yao, J.; Chen, K. Selective Flotation Separation of Apatite from Dolomite Utilizing a Novel Eco-Friendly and Efficient Depressant for Sustainable Manufacturing of Phosphate Fertilizer. *J. Clean. Prod.* **2021**, *286*, 124949. [CrossRef]
33. George, P.; Nguyen, A.V.; Jameson, G.J. Assessment of True Flotation and Entrainment in the Flotation of Submicron Particles by Fine Bubbles. *Miner. Eng.* **2004**, *17*, 847–853. [CrossRef]
34. Farrokhpay, S.; Ndlovu, B.; Bradshaw, D. Behaviour of Swelling Clays versus Non-Swelling Clays in Flotation. *Miner. Eng.* **2016**, *96–97*, 59–66. [CrossRef]
35. Verdugo, L.; Zhang, L.; Etschmann, B.; Brugger, J.; Bruckard, W.; Menacho, J.; Molina, L.; Hoadley, A. Spent Lithium-Ion Battery Recycling Using Flotation Technology: Effect of Material Heterogeneity on Separation Performance. *Processes* **2024**, *12*, 1363. [CrossRef]
36. Abdollahi, S.; Afraei, S.; Mehdilo, A.; Kouchakzadeh, R.; Irannajad, M. A New Mixture of Anionic Collectors for Improvement of Apatite Floatability. *Miner. Eng.* **2025**, *222*, 109129. [CrossRef]
37. Wills, B.A.; Finch, J.A. Chapter 12—Froth Flotation. In *Wills' Mineral Processing Technology*, 8th ed.; Wills, B.A., Finch, J.A., Eds.; Butterworth-Heinemann: Boston, UK, 2016; pp. 265–380. ISBN 978-0-08-097053-0.
38. Pasero, M.; Kampf, A.R.; Ferraris, C.; Pekov, I.V.; Rakovan, J.; White, T.J. Nomenclature of the Apatite Supergroup Minerals. *Eur. J. Mineral.* **2010**, *22*, 163–179. [CrossRef]
39. Bulina, N.V.; Makarova, S.V.; Prosanov, I.Y.; Vinokurova, O.B.; Lyakhov, N.Z. Structure and Thermal Stability of Fluorhydroxyapatite and Fluorapatite Obtained by Mechanochemical Method. *J. Solid State Chem.* **2020**, *282*, 121076. [CrossRef]
40. Simukanga, S.; Lombe, W.C. Electrochemical Properties of Apatite and Other Minerals of Zambian Phosphate Ores in Aqueous Solution. *Fertil. Res.* **1995**, *41*, 159–166. [CrossRef]
41. Engvik, A.K.; Golla-Schindler, U.; Berndt, J.; Austrheim, H.; Putnis, A. Intragranular Replacement of Chlorapatite by Hydroxy-Fluor-Apatite during Metasomatism. *Lithos* **2009**, *112*, 236–246. [CrossRef]
42. Rulis, P.; Yao, H.; Ouyang, L.; Ching, W.Y. Electronic Structure, Bonding, Charge Distribution, and x-Ray Absorption Spectra of the (001) Surfaces of Fluorapatite and Hydroxyapatite from First Principles. *Phys. Rev. B* **2007**, *76*, 245410. [CrossRef]
43. Xie, J.; Li, X.; Mao, S.; Li, L.; Ke, B.; Zhang, Q. Effects of Structure of Fatty Acid Collectors on the Adsorption of Fluorapatite (0 0 1) Surface: A First-Principles Calculations. *Appl. Surf. Sci.* **2018**, *444*, 699–709. [CrossRef]
44. Comodi, P.; Liu, Y.; Zanazzi, P.F.; Montagnoli, M. Structural and Vibrational Behaviour of Fluorapatite with Pressure. Part I: In Situ Single-Crystal X-Ray Diffraction Investigation. *Phys. Chem. Miner.* **2001**, *28*, 219–224. [CrossRef]
45. Kim, J.Y.; Fenton, R.R.; Hunter, B.A.; Kennedy, B.J. Powder Diffraction Studies of Synthetic Calcium and Lead Apatites. *Aust. J. Chem.* **2000**, *53*, 679–686. [CrossRef]
46. Fuerstenau, D.W. Pradip Zeta Potentials in the Flotation of Oxide and Silicate Minerals. *Adv. Colloid Interface Sci.* **2005**, *114–115*, 9–26. [CrossRef] [PubMed]
47. Kholodov, V.N.; Nedumov, R.I. The Role of Black Shales in the Formation of Phosphate and Manganese Ores. *Lithol. Miner. Resour.* **2011**, *46*, 321–352. [CrossRef]
48. Liu, S.; Zhang, W.; Ren, Q.; Tu, R.; Qiu, F.; Gao, Z.; Xu, S.; Sun, W.; Tian, M. Innovating an Amphoteric Collector Derived from Dodecylamine Molecular Structure: Facilitating the Selective Flotation Separation of Apatite from Quartz and Dolomite. *Miner. Eng.* **2024**, *215*, 108819. [CrossRef]
49. Derhy, M.; Taha, Y.; Ait-khouia, Y.; Elghali, A.; Benzaazoua, M.; Hakkou, R. Enhancing Selective Calcite and Dolomite Flotation in the Phosphate Ores: Investigation, Modeling, and Automated Mineralogy Assessment. *Miner. Eng.* **2024**, *207*, 108569. [CrossRef]
50. Lan, S.; Shen, P.; Zheng, Q.; Qiao, L.; Dong, L.; Liu, D. Effective Flotation Separation of Apatite from Dolomite Using a New Eco-Friendly Depressant Gallic acid. *Green Chem.* **2024**, *26*, 1627–1636. [CrossRef]
51. Krasowska, M.; Zawala, J.; Bradshaw-Hajek, B.H.; Ferri, J.K.; Beattie, D.A. Interfacial Characterisation for Flotation: 1. Solid-Liquid Interface. *Curr. Opin. Colloid Interface Sci.* **2018**, *37*, 61–73. [CrossRef]

52. Derhy, M.; Taha, Y.; El-Bahi, A.; Ait-Khouia, Y.; Benzaazoua, M.; Hakkou, R. Selective Flotation of Calcite and Dolomite from Apatite Using Bio-Based Alternatives to Conventional Collectors: Castor and Mustard Oils. *Miner. Eng.* **2024**, *208*, 108597. [CrossRef]
53. Liu, W.; Ge, R.; Guo, Y.; Xia, Y.; Zhao, S.; Liu, W.; Cui, B.; Shen, Y. Study on the Performance and Mechanism of Novel Hydroxyl Quaternary Ammonium Collector for Separation of Quartz from Apatite. *J. Mol. Liq.* **2025**, *421*, 126868. [CrossRef]
54. Peng, X.; Liu, W.; Zhao, Q.; Liu, W.; Tong, K.; Zhao, P. Development and Utilization of a Novel Hydrogen Bonding Enhanced Collector in the Separation of Apatite from Quartz. *Miner. Eng.* **2022**, *180*, 107477. [CrossRef]
55. Derhy, M.; Taha, Y.; Hakkou, R.; Benzaazoua, M. Review of the Main Factors Affecting the Flotation of Phosphate Ores. *Minerals* **2020**, *10*, 1109. [CrossRef]
56. Ulusoy, U.; Yekeler, M. Correlation of the Surface Roughness of Some Industrial Minerals with Their Wettability Parameters. *Chem. Eng. Process. Process Intensif.* **2005**, *44*, 555–563. [CrossRef]
57. Wang, X.; Zhang, Q. Insight into the Influence of Surface Roughness on the Wettability of Apatite and Dolomite. *Minerals* **2020**, *10*, 114. [CrossRef]
58. Jiang, C.; Chen, H.; Li, S.; Cao, Y.; Ao, X. Effect of Mg, Al, and Fe Impurities on the Wettability of the Fluorapatite (001) Surface. *Colloids Surf. A Physicochem. Eng. Asp.* **2022**, *654*, 130130. [CrossRef]
59. Gao, Z.-Y.; Jiang, Z.-Y.; Sun, W.; Gao, Y.-S. Typical Roles of Metal Ions in Mineral Flotation: A Review. *Trans. Nonferrous Met. Soc. China* **2021**, *31*, 2081–2101. [CrossRef]
60. Parapari, P.S.; Irannajad, M.; Mehdilo, A. Modification of Ilmenite Surface Properties by Superficial Dissolution Method. *Miner. Eng.* **2016**, *92*, 160–167. [CrossRef]
61. Zhang, H.; Zhou, F.; Yu, H.; Liu, M. Double Roles of Sodium Hexametaphosphate in the Flotation of Dolomite from Apatite. *Colloids Surf. A Physicochem. Eng. Asp.* **2021**, *626*, 127080. [CrossRef]
62. Abdel-Khalek, N.A. Evaluation of Flotation Strategies for Sedimentary Phosphates with Siliceous and Carbonates Gangues. *Miner. Eng.* **2000**, *13*, 789–793. [CrossRef]
63. Sun, R.; Xing, D.; Deng, J.; Jin, D.; Ma, S.; Liu, J.; Li, G.; Qin, G.; Liu, X. Effect of Sulfuric Acid Pretreatment on Flotation Performance of Calcite and Fluorite. *Miner. Eng.* **2023**, *203*, 108301. [CrossRef]
64. Chen, J.; Yuan, X.; Yin, Y.; Wang, M.; Cheng, D.; Ao, X.; Xie, Y. Effect of Ca2+ Dissolution and Migration Transformation from Phosphorite on the Surface Properties of Dolomite. *Miner. Eng.* **2022**, *188*, 107824. [CrossRef]
65. Németh, Z.; Csóka, I.; Semnani Jazani, R.; Sipos, B.; Haspel, H.; Kozma, G.; Kónya, Z.; Dobó, D.G. Quality by Design-Driven Zeta Potential Optimisation Study of Liposomes with Charge Imparting Membrane Additives. *Pharmaceutics* **2022**, *14*, 1798. [CrossRef] [PubMed]
66. Kou, J.; Tao, D.; Xu, G. Fatty Acid Collectors for Phosphate Flotation and Their Adsorption Behavior Using QCM-D. *Int. J. Miner. Process.* **2010**, *95*, 1–9. [CrossRef]
67. Cheng, R.; Li, C.; Liu, X.; Deng, S. Synergism of Octane Phenol Polyoxyethylene-10 and Oleic Acid in Apatite Flotation. *Physicochem. Probl. Miner. Process.* **2017**, *53*, 1214–1227. [CrossRef]
68. Perrone, J.; Fourest, B.; Giffaut, E. Surface Characterization of Synthetic and Mineral Carbonate Fluoroapatites. *J. Colloid Interface Sci.* **2002**, *249*, 441–452. [CrossRef]
69. Chaïrat, C.; Oelkers, E.H.; Schott, J.; Lartigue, J.-E. Fluorapatite Surface Composition in Aqueous Solution Deduced from Potentiometric, Electrokinetic, and Solubility Measurements, and Spectroscopic Observations. *Geochim. Cosmochim. Acta* **2007**, *71*, 5888–5900. [CrossRef]
70. Yuehua, H.; Chi, R.; Xu, Z. Solution Chemistry Study of Salt-Type Mineral Flotation Systems: Role of Inorganic Dispersants. *Ind. Eng. Chem. Res.* **2003**, *42*, 1641–1647. [CrossRef]
71. Nduwa-Mushidi, J.; Anderson, C.G. Surface Chemistry and Flotation Behaviors of Monazite–Apatite–Ilmenite–Quartz–Rutile–Zircon with Octanohydroxamic Acid. *J. Sustain. Metall.* **2017**, *3*, 62–72. [CrossRef]
72. Amankonah, J.O.; Somasundaran, P. Effects of Dissolved Mineral Species on the Electrokinetic Behavior of Calcite and Apatite. *Colloids Surf.* **1985**, *15*, 335–353. [CrossRef]
73. Mishra, S.K. The Electrokinetics of Apatite and Calcite in Inorganic Electrolyte Environment. *Int. J. Miner. Process.* **1978**, *5*, 69–83. [CrossRef]
74. Dong, L.; Jiao, F.; Qin, W.; Zhu, H.; Jia, W. Selective Depressive Effect of Sodium Fluorosilicate on Calcite during Scheelite Flotation. *Miner. Eng.* **2019**, *131*, 262–271. [CrossRef]
75. Angélica Evangelista de Carvalho, J.; Roberto Gomes Brandão, P.; Bicalho Henriques, A.; Silva de Oliveira, P.; Zanoni Lopes Cançado, R.; Rodrigues da Silva, G. Selective Flotation of Apatite from Micaceous Minerals Using Patauá Palm Tree Oil Collector. *Miner. Eng.* **2020**, *156*, 106474. [CrossRef]
76. Santos, L.H.; Otávio dos Santos, G.; Fernandes de Magalhães, L.; Rodrigues da Silva, G.; Eduardo Clark Peres, A. Application of Andiroba Oil as a Novel Collector in Apatite Flotation. *Miner. Eng.* **2022**, *185*, 107678. [CrossRef]

77. Abdel-Halim, M.M.; Abdel Khalek, M.A.; Zheng, R.; Gao, Z. Sodium N-Lauroylsarcosinate (SNLS) as a Selective Collector for Calcareous Phosphate Beneficiation. *Minerals* **2022**, *12*, 829. [CrossRef]
78. Liu, W.; Zhang, S.; Wang, W.; Zhang, J.; Yan, W.; Deng, J.; Feng, Q.; Huang, Y. The Effects of Ca(II) and Mg(II) Ions on the Flotation of Spodumene Using NaOL. *Miner. Eng.* **2015**, *79*, 40–46. [CrossRef]
79. Castro, S.; Lopez-Valdivieso, A.; Laskowski, J.S. Review of the Flotation of Molybdenite. Part I: Surface Properties and Floatability. *Int. J. Miner. Process.* **2016**, *148*, 48–58. [CrossRef]
80. Liao, S.; Wu, Y.; Yang, Y.; Wu, L.; Gu, G.; Wang, Y. Flotation Behavior and Mechanism of Microfine Apatite and Illite in Different Electrolyte Solution Systems. *Colloids Surf. A Physicochem. Eng. Asp.* **2024**, *699*, 134563. [CrossRef]
81. Rao, K.H.; Antti, B.-M.; Forssberg, E. Mechanism of Oleate Interaction on Salt-Type Minerals, Part II. Adsorption and Electrokinetic Studies of Apatite in the Presence of Sodium Oleate and Sodium Metasilicate. *Int. J. Miner. Process.* **1990**, *28*, 59–79. [CrossRef]
82. Jong, K.; Han, Y.; Ryom, S. Flotation Mechanism of Oleic Acid Amide on Apatite. *Colloids Surf. A Physicochem. Eng. Asp.* **2017**, *523*, 127–131. [CrossRef]
83. de Oliveira, P.; Mansur, H.; Mansur, A.; da Silva, G.; Clark Peres, A.E. Apatite Flotation Using Pataua Palm Tree Oil as Collector. *J. Mater. Res. Technol.* **2019**, *8*, 4612–4619. [CrossRef]
84. Xu, W.; Mei, G.; Tian, Y.; Shi, B.; Guo, C.; Pan, W. Reverse Cationic Flotation of Low-Grade Phosphate Ore Using Quaternary Ammonium Salt as a Collector and Its Adsorption Mechanism. *Green Smart Min. Eng.* **2024**, *1*, 322–335. [CrossRef]
85. El-bahi, A.; Taha, Y.; Ait-Khouia, Y.; Elghali, A.; Benzaazoua, M. Enhancing Sustainability in Phosphate Ore Processing: Performance of Frying Oil as Alternative Flotation Collector for Carbonate Removal. *Int. J. Min. Sci. Technol.* **2024**, *34*, 557–571. [CrossRef]
86. Ding, Z.; Li, J.; Bi, Y.; Yu, P.; Dai, H.; Wen, S.; Bai, S. The Adsorption Mechanism of Synergic Reagents and Its Effect on Apatite Flotation in Oleamide-Sodium Dodecyl Benzene Sulfonate (SDBS) System. *Miner. Eng.* **2021**, *170*, 107070. [CrossRef]
87. Wang, X. Surface Wettability Regulation of Minerals in Flotation of Calcium-Magnesium Phosphate Ore. Ph.D. Thesis, Guizhou University, Guiyang, China, 2022.
88. Budemberg, G.; Jolsterå, R.; Chelgani, S.C. Eco-Friendly Collectors in Apatite Froth Flotation: A Review. *Int. J. Min. Sci. Technol.* **2025**, *35*, 539–551. [CrossRef]
89. Lan, S.; Dong, L.; Shen, P.; Zheng, Q.; Qiao, L.; Liu, D. Synergistic Impact of Surfactant and Sodium Oleate on Dolomite Removal from Fluorapatite via Reverse Flotation. *Appl. Surf. Sci.* **2024**, *655*, 159504. [CrossRef]
90. Zhang, J.; Li, H.; Chai, Y.; Zhang, Q. Crystal Properties and Interaction with Flotation Reagent of Fluorapatite and Dolomite. *Miner. Eng.* **2023**, *201*, 108204. [CrossRef]
91. Wang, X.; Zhang, Q.; Chen, J.; Xie, J.; Xie, F.; Li, L. Electronic Properties and Amine Collectors Effect of Fluorapatite and Quartz Surface. *J. Guizhou Univ.* **2017**, *34*, 21–28.
92. Dederichs, T.; Möller, M.; Weichold, O. Colloidal Stability of Hydrophobic Nanoparticles in Ionic Surfactant Solutions: Definition of the Critical Dispersion Concentration. *Langmuir* **2009**, *25*, 2007–2012. [CrossRef] [PubMed]
93. Veeramanoharan, A.; Kim, S.-C. A Comprehensive Review on Sustainable Surfactants from CNSL: Chemistry, Key Applications and Research Perspectives. *RSC Adv.* **2024**, *14*, 25429–25471. [CrossRef]
94. Zhang, W.; Tu, R.; Ren, Q.; Liu, S.; Guo, Z.; Liu, P.; Tian, M. The Selective Flotation Separation of Apatite from Dolomite and Calcite Utilizing a Combination of 2-Chloro-9-Octadecenoic Acid Collector and Sodium Pyrophosphate Depressant. *Sep. Purif. Technol.* **2025**, *353*, 128446. [CrossRef]
95. Abdel-Halim, M.M.; Fan, R.; Khalek, M.A.A.; Zheng, R.; Xu, S.; Gao, Z. Apatite–Calcite Flotation Separation Using Sodium N-Lauroylsarcosinate as a Selective Collector. *Minerals* **2023**, *13*, 970. [CrossRef]
96. Ruan, Y.; Deng, B.; He, D.; Chi, R. Synergetic Effect of Cottonseed Fatty Acid Salt and Nonionic Surfactant NP-4 in the Froth Flotation of Siliceous-Calcareous Phosphate Rock. *Colloids Surf. A Physicochem. Eng. Asp.* **2021**, *622*, 126650. [CrossRef]
97. Wang, L.; Tian, M.; Khoso, S.A.; Hu, Y.; Sun, W.; Gao, Z. Improved Flotation Separation of Apatite from Calcite with Benzohydroxamic Acid Collector. *Miner. Process. Extr. Metall. Rev.* **2019**, *40*, 427–436. [CrossRef]
98. Liu, W.; Liu, W.; Wang, B.; Duan, H.; Peng, X.; Chen, X.; Zhao, Q. Novel Hydroxy Polyamine Surfactant N-(2-Hydroxyethyl)-N-Dodecyl-Ethanediamine: Its Synthesis and Flotation Performance Study to Quartz. *Miner. Eng.* **2019**, *142*, 105894. [CrossRef]
99. Wang, B.; Zhao, P.; Liu, W.; Liu, W.; Guo, Y.; Tong, K.; Chen, X. Novel Hydroxyl-Containing Quaternary Ammonium Salt N-(2-Hydroxyethyl)-N, N-Dimethyl-3-[(1-Oxododecyl)Amino]-1-Propanaminium: Its Synthesis and Flotation Performance to Quartz. *Minerals* **2023**, *13*, 702. [CrossRef]
100. Farid, Z.; Assimeddine, M.; Abdennouri, M.; Barka, N.; Sadiq, M. Lecithin as an Ecofriendly Amphoteric Collector Foaming Agent for the Reverse Flotation of Phosphate Ore. *Chem. Eng. Res. Des.* **2023**, *200*, 292–302. [CrossRef]
101. Li, J.; Nie, G.; Li, J.; Zhu, Z.; Wang, Z. Flotation Separation of Quartz and Dolomite from Collophane Using Sodium N-Dodecyl-β-Amino Propionate and Its Adsorption Mechanism. *Colloids Surf. A Physicochem. Eng. Asp.* **2022**, *641*, 128586. [CrossRef]
102. Filippov, L.O.; Filippova, I.V.; Ardian, R.; Fornasiero, D. Ethoxylated Alcohols as Co-Collectors in Apatite, Calcite and Dolomite Flotation. *Miner. Eng.* **2025**, *224*, 109173. [CrossRef]

103. Severov, V.V.; Filippova, I.V.; Filippov, L.O. Use of Fatty Acids with an Ethoxylated Alcohol for Apatite Flotation from Old Fine-Grained Tailings. *Miner. Eng.* **2022**, *188*, 107832. [CrossRef]
104. Isah, A.; Arif, M.; Hassan, A.; Mahmoud, M.; Iglauer, S. Fluid–Rock Interactions and Its Implications on EOR: Critical Analysis, Experimental Techniques and Knowledge Gaps. *Energy Rep.* **2022**, *8*, 6355–6395. [CrossRef]
105. Aarab, I.; Derqaoui, M.; Abidi, A.; Yaacoubi, A.; El Amari, K.; Etahiri, A.; Baçaoui, A. Direct Flotation of Low-Grade Moroccan Phosphate Ores: A Preliminary Micro-Flotation Study to Develop New Beneficiation Routes. *Arab. J. Geosci.* **2020**, *13*, 1252. [CrossRef]
106. Gong, W.Q.; Parentich, A.; Little, L.H.; Warren, L.J. Adsorption of Oleate on Apatite Studied by Diffuse Reflectance Infrared Fourier Transform Spectroscopy. *Langmuir* **1992**, *8*, 118–124. [CrossRef]
107. Li, X.; Zhang, Q.; Ke, B.; Mao, S. Density Functional Theory of Molecular Structure and Properties of Fatty Acid Collectors. *Multipurp. Util. Miner. Resour.* **2024**, *45*, 167–173. [CrossRef]
108. Kang, Y.; Zhang, Q. The Flotation Separation of Apatite from Dolomite Using a Fatty Acid-Based Collector and the Mechanisms of Adsorption. *Surf. Interface Anal.* **2023**, *55*, 138–150. [CrossRef]
109. Fuerstenau, M.C. Chapter 8—Flotation. In *Principles of Mineral Processing*; Society for Mining, Metallurgy, and Exploration, Inc.: Littleton, CO, USA, 2003; p. 252. ISBN 978-0-87335-167-6.
110. Cook, B.K.; Gibson, C.E. A Review of Fatty Acid Collectors: Implications for Spodumene Flotation. *Minerals* **2023**, *13*, 212. [CrossRef]
111. Houot, R. Beneficiation of Phosphatic Ores through Flotation: Review of Industrial Applications and Potential Developments. *Int. J. Miner. Process.* **1982**, *9*, 353–384. [CrossRef]
112. de Lima, F.V.; Budemberg, G.; Cruz, S.H.; Braga, A.S. A Collector Promoter for Apatite Flotation in the Serra Do Salitre Complex. *Minerals* **2023**, *13*, 599. [CrossRef]
113. Sis, H.; Chander, S. Improving Froth Characteristics and Flotation Recovery of Phosphate Ores with Nonionic Surfactants. *Miner. Eng.* **2003**, *16*, 587–595. [CrossRef]
114. Giesekke, E.W.; Harris, P.J. The Role of Polyoxyethylene Alkyl Ethers in Apatite Flotation at Foskor, Phalaborwa (South Africa). *Miner. Eng.* **1994**, *7*, 1345–1361. [CrossRef]
115. Cao, Q.; Cheng, J.; Wen, S.; Li, C.; Liu, J. Synergistic Effect of Dodecyl Sulfonate on Apatite Flotation with Fatty Acid Collector. *Sep. Sci. Technol.* **2016**, *51*, 1389–1396. [CrossRef]
116. Du, W.; Li, X.; Zhang, Q. DFT Study of Coadsorption of Fatty Acid and Kerosene on Fluorapatite (001) Surface. *Physicochem. Probl. Miner. Process.* **2023**, *59*, 161890. [CrossRef]
117. Yu, H.; Zhu, Y.; Lu, L.; Hu, X.; Li, S. Removal of Dolomite and Potassium Feldspar from Apatite Using Simultaneous Flotation with a Mixed Cationic-Anionic Collector. *Int. J. Min. Sci. Technol.* **2023**, *33*, 783–791. [CrossRef]
118. Yang, L.; Li, X.; Li, W.; Yan, X.; Zhang, H. Intensification of Interfacial Adsorption of Dodecylamine onto Quartz by Ultrasonic Method. *Sep. Purif. Technol.* **2019**, *227*, 115701. [CrossRef]
119. Zhang, W.; Liu, S.; Ren, Q.; Tu, R.; Qiu, F.; Xu, S.; Sun, W.; Tian, M. Proposing a Novel Factor Influencing Flotation Collector Abilities to Collect Minerals by Comparing Two Cationic Collectors N,N-Bis(2-Hydroxy-3-Chloropropyl) Dodecylamine and Dodecylamine. *Appl. Surf. Sci.* **2023**, *640*, 158284. [CrossRef]
120. Novich, B.E.; Ring, T.A. Predictive Model for The Alkylamine-Quartz Flotation System. *Langmuir* **1985**, *1*, 701–708. [CrossRef]
121. Zhang, W.; Wang, Z.; Tian, H.; Liu, J.; Guo, Z.; Liu, P.; Tian, M.; Xie, L. Experimental Evidence Supporting Distinct Adsorption Configurations of N,N-Bis(2-hydroxy-3-Chloropropyl) Dodecylamine and Dodecylamine on Apatite Surfaces: Quartz Crystal Microbalance and Atomic Force Microscopy Studies. *Surf. Interfaces* **2024**, *51*, 104681. [CrossRef]
122. Luo, B.; Zhu, Y.; Sun, C.; Li, Y.; Han, Y. Flotation and Adsorption of a New Collector α-Bromodecanoic Acid on Quartz Surface. *Miner. Eng.* **2015**, *77*, 86–92. [CrossRef]
123. Sun, H.; Yang, B.; Zhu, Z.; Yin, W.; Sheng, Q.; Hou, Y.; Yao, J. New Insights into Selective-Depression Mechanism of Novel Depressant EDTMPS on Magnesite and Quartz Surfaces: Adsorption Mechanism, DFT Calculations, and Adsorption Model. *Miner. Eng.* **2021**, *160*, 106660. [CrossRef]
124. Liang, Q.; Xu, W.; Mei, G.; Tian, Y.; Guo, C.; Pan, W. Synthesis of a Photosensitive Quaternary Ammonium Collector and Its Flotation Performance and Mechanism for Quartz. *Colloids Surf. A Physicochem. Eng. Asp.* **2023**, *661*, 130936. [CrossRef]
125. Rath, S.S.; Sahoo, H.; Das, B.; Mishra, B.K. Density Functional Calculations of Amines on the (101) Face of Quartz. *Miner. Eng.* **2014**, *69*, 57–64. [CrossRef]
126. Zhang, W.; Ren, Q.; Tu, R.; Liu, S.; Qiu, F.; Guo, Z.; Liu, P.; Xu, S.; Sun, W.; Tian, M. The Application of a Novel Amine Collector, 1-(Dodecylamino)-2-Propanol, in the Reverse Flotation Separation of Apatite and Quartz. *J. Mol. Liq.* **2024**, *399*, 124377. [CrossRef]
127. Shao, X.; Jiang, C.L.; Parekh, B.K. Enhanced Flotation Separation of Phosphate and Dolomite Using a New Amphoteric Collector. *Min. Metall. Explor.* **1998**, *15*, 11–14. [CrossRef]
128. Somasundaran, P.; Zhang, L. Adsorption of Surfactants on Minerals for Wettability Control in Improved Oil Recovery Processes. *J. Pet. Sci. Eng.* **2006**, *52*, 198–212. [CrossRef]

129. Pugh, R.J. The Role of the Solution Chemistry of Dodecylamine and Oleic Acid Collectors in the Flotation of Fluorite. *Colloids Surf.* **1986**, *18*, 19–41. [CrossRef]
130. Rao, C.Y.V. Magnetically Coupled Sealless Pumps. *Chem. Eng. World* **1997**, *32*, 67–69.
131. Filippov, L.O.; Filippova, I.V.; Lafhaj, Z.; Fornasiero, D. The Role of a Fatty Alcohol in Improving Calcium Minerals Flotation with Oleate. *Colloids Surf. A Physicochem. Eng. Asp.* **2019**, *560*, 410–417. [CrossRef]
132. Liu, S.; Han, B.; Zhao, T. The Effect of Various Surfactants on Fatty Acid for Apatite Flotation and Their Adsorption Mechanizm. *Physicochem. Probl. Miner. Process.* **2021**, *57*, 46–56.
133. Pawliszak, P.; Beheshti, A.; Møller, A.; Blencowe, A.; Beattie, D.A.; Krasowska, M. Increasing Surface Hydrophilicity with Biopolymers: A Combined Single Bubble Collision, QCM-D and AFM Study. *J. Colloid Interface Sci.* **2024**, *667*, 393–402. [CrossRef]
134. Yu, L.; Yu, P.; Bai, S. A Critical Review on the Flotation Reagents for Phosphate Ore Beneficiation. *Minerals* **2024**, *14*, 828. [CrossRef]
135. Liu, X.; Ruan, Y.; Li, C.; Cheng, R. Effect and Mechanism of Phosphoric Acid in the Apatite/Dolomite Flotation System. *Int. J. Miner. Process.* **2017**, *167*, 95–102. [CrossRef]
136. Al-Fariss, T.F.; Arafat, Y.; Abd El-Aleem, F.A.; El-Midany, A.A. Investigating Sodium Sulphate as a Phosphate Depressant in Acidic Media. *Sep. Purif. Technol.* **2014**, *124*, 163–169. [CrossRef]
137. Oulkhir, A.; Lyamlouli, K.; Oussfan, A.; Orange, F.; Etahiri, A.; Benhida, R. Efficient Flotation Separation Approach of Apatite from Calcite for Phosphate Up-Grading Using Phosphorylated Starch Macromolecules as a Selective Depressant. *Carbohydr. Polym.* **2025**, *348*, 122878. [CrossRef] [PubMed]
138. Dai, L.; Feng, B.; Zhang, W.; Guo, W. Study of the Effect and Mechanism of Environmentally Friendly Depressant Tragacanth Gum on Scheelite and Apatite Separation. *Process Saf. Environ. Prot.* **2025**, *196*, 106913. [CrossRef]
139. Jing, L.; Xu, L.; Xue, K.; Wang, D.; Ma, Z.; Meng, J.; Shi, X.; Liu, C. Selective Depression by Using Environment-Friendly Depressant Pectin in Apatite and Dolomite Flotation System. *Miner. Eng.* **2023**, *203*, 108373. [CrossRef]
140. Zhong, C.; Feng, B.; Zhang, L.; Zhang, W.; Wang, H.; Gao, Z. Flotation Separation of Apatite and Calcite Using Gum Arabic as a Depressant. *Colloids Surf. A Physicochem. Eng. Asp.* **2022**, *632*, 127723. [CrossRef]
141. Wang, T.; Feng, B.; Guo, Y.; Zhang, W.; Rao, Y.; Zhong, C.; Zhang, L.; Cheng, C.; Wang, H.; Luo, X. The Flotation Separation Behavior of Apatite from Calcite Using Carboxymethyl Chitosan as Depressant. *Miner. Eng.* **2020**, *159*, 106635. [CrossRef]
142. Jing, L.; Xue, K.; Tian, J.; Zhang, X.; Wang, D.; Guo, W.; Ma, Z.; Xu, L. Separation Mechanism of Apatite and Dolomite with Flotation Depressant Konjac Glucomannan. *Miner. Eng.* **2024**, *216*, 108844. [CrossRef]
143. Wang, Y.; Ahmed Khoso, S.; Luo, X.; Tian, M. Understanding the Depression Mechanism of Citric Acid in Sodium Oleate Flotation of Ca^{2+}-Activated Quartz: Experimental and DFT Study. *Miner. Eng.* **2019**, *140*, 105878. [CrossRef]
144. Dho, H.; Iwasaki, I. Role of Sodium Silicate in Phosphate Flotation. *Min. Metall. Explor.* **1990**, *7*, 215–221. [CrossRef]
145. Zheng, X.; Arps, P.J.; Smith, R.W. Adhesion of Two Bacteria onto Dolomite and Apatite: Their Effect on Dolomite Depression in Anionic Flotation. *Int. J. Miner. Process.* **2001**, *62*, 159–172. [CrossRef]
146. Somasundaran, P.; Wang, D. *Solution Chemistry: Minerals and Reagents*, 1st ed.; Developments in Mineral Processing; Elsevier: Amsterdam, The Netherlands, 2006; Volume 17, ISBN 978-0-444-52059-3.
147. Prasad, M.; Majumder, A.K.; Rao, T.C. Reverse Flotation of Sedimentary Calcareous/Dolomitic Rock Phosphate Ore—An Overview. *Min. Metall. Explor.* **2000**, *17*, 49–55. [CrossRef]
148. Filippov, L.O.; Filippova, I.V.; Kaba, O.B.; Fornasiero, D. In-Situ Study of the Kinetics of Phosphoric Acid Interaction with Calcite and Fluorapatite by Raman Spectroscopy and Flotation. *Miner. Eng.* **2021**, *162*, 106729. [CrossRef]
149. Chen, Y.; Wu, J.; Liu, R.; Liu, C.; Liu, L.; Li, R.; Zhang, H.; Pang, J.; Liu, D. Application of Waste Acid from Phosphogysum Dam as an Eco-Friendly Depressant in Collophane Flotation. *J. Clean. Prod.* **2020**, *267*, 122184. [CrossRef]
150. Liu, X.; Li, C.; Luo, H.; Cheng, R.; Liu, F. Selective Reverse Flotation of Apatite from Dolomite in Collophanite Ore Using Saponified Gutter Oil Fatty Acid as a Collector. *Int. J. Miner. Process.* **2017**, *165*, 20–27. [CrossRef]
151. Chen, Y.; Feng, Q.; Zhang, G.; Liu, D.; Liu, R. Effect of Sodium Pyrophosphate on the Reverse Flotation of Dolomite from Apatite. *Minerals* **2018**, *8*, 278. [CrossRef]
152. Elgillani, D.A.; Abouzeid, A.-Z.M. Flotation of Carbonates from Phosphate Ores in Acidic Media. *Int. J. Miner. Process.* **1993**, *38*, 235–256. [CrossRef]
153. Nagaraj, D.R.; Rothenberg, A.S.; Lipp, D.W.; Panzer, H.P. Low Molecular Weight Polyacrylamide-Based Polymers as Modifiers in Phosphate Beneficiation. *Int. J. Miner. Process.* **1987**, *20*, 291–308. [CrossRef]
154. Wang, X.; Song, Q.; Xie, R.; Liu, J.; Zhu, Y. Selective Flotation Separation of Scheelite from Apatite by Application of ATMP as an Efficient Depressant. *J. Mol. Liq.* **2023**, *378*, 121604. [CrossRef]
155. Zhang, P.; Snow, R. Evaluation of Phosphate Depressants in the Phosphate/Silica System. *Min. Metall. Explor.* **2009**, *26*, 101–104. [CrossRef]
156. Chen, W.; Feng, Q.; Zhang, G.; Yang, Q.; Zhang, C. The Effect of Sodium Alginate on the Flotation Separation of Scheelite from Calcite and Fluorite. *Miner. Eng.* **2017**, *113*, 1–7. [CrossRef]

157. Zhong, C.; Wang, H.; Feng, B.; Zhang, L.; Chen, Y.; Gao, Z. Flotation Separation of Scheelite and Apatite by Polysaccharide Depressant Xanthan Gum. *Miner. Eng.* **2021**, *170*, 107045. [CrossRef]
158. Wang, L.; Lyu, W.; Li, F.; Liu, J.; Zhang, H. Discrepant Adsorption Behavior of Sodium Alginate onto Apatite and Calcite Surfaces: Implications for Their Selective Flotation Separation. *Miner. Eng.* **2022**, *181*, 107553. [CrossRef]
159. McLean, R.J.C.; Beauchemin, D.; Clapham, L.; Beveridge, T.J. Metal-Binding Characteristics of the Gamma-Glutamyl Capsular Polymer of Bacillus Licheniformis ATCC 9945. *Appl. Environ. Microbiol.* **1990**, *56*, 3671–3677. [CrossRef] [PubMed]
160. Li, J.X.; Nie, G.H.; Jiang, Y.; Luo, G.J.; Li, J. Selective Depression of Escherichia Coli on Flotation of Collophanite and Dolomite. *Physicochem. Probl. Miner. Process.* **2022**, *58*, 150604. [CrossRef]
161. Rabia, H.; Ould Hamou, M.; Kasperkiewicz, K.; Brożek, J.; Augustyniak, M. Adhesion Abilities and Biosorption of Cd and Mg by Microorganisms—First Step for Eco-Friendly Beneficiation of Phosphate Ore. *Sci. Rep.* **2019**, *9*, 12929. [CrossRef] [PubMed]
162. Qun, W.; Heiskanen, K. Batch Flotation Tests by Fatty Acid on a Phosphate-Iron Oxide-Silicate Regolith Ore Sample from Sokli, Finland. *Miner. Eng.* **1990**, *3*, 473–481. [CrossRef]
163. Sayilgan, A.; Arol, A.I. Effect of Carbonate Alkalinity on Flotation Behavior of Quartz. *Int. J. Miner. Process.* **2004**, *74*, 233–238. [CrossRef]
164. Wang, L.; Gao, H.; Song, S.; Zhou, W.; Xue, N.; Nie, Y.; Feng, B. The Depressing Role of Sodium Alginate in the Flotation of Ca^{2+}-Activated Quartz Using Fatty Acid Collector. *J. Mol. Liq.* **2021**, *343*, 117618. [CrossRef]
165. Dai, L.; Feng, B.; Zhang, L.; Chen, Y.; Bayoundoula, J. Selective Flotation Separation of Spodumene and Quartz with Carboxylated Chitosan as Depressant. *Miner. Eng.* **2023**, *203*, 108343. [CrossRef]
166. Xiang, Z.; Feng, B.; Zhang, L.; Bayoundoula, J.; Wang, Z. The Role of Depressant Pectin in the Flotation Separation of Spodumene and Quartz. *Miner. Eng.* **2023**, *203*, 108320. [CrossRef]
167. Zhu, Y.; Yang, L.; Hu, X.; Zhang, X.; Zheng, G. Flotation Separation of Quartz from Magnesite Using Carboxymethyl Cellulose as Depressant. *Trans. Nonferrous Met. Soc. China* **2022**, *32*, 1623–1637. [CrossRef]
168. Sun, N.; Wang, G.; Ge, P.; Sun, W.; Xu, L.; Tang, H.; Wang, L. Selective Flotation of Quartz from Feldspar Using Hydroxypropyl Starch as Depressant. *Miner. Eng.* **2023**, *195*, 108022. [CrossRef]
169. Sokolović, J.; Mišković, S. The Effect of Particle Size on Coal Flotation Kinetics: A Review. *Physicochem. Probl. Miner. Process.* **2018**, *54*, 1172–1190.
170. Zhang, D.; Ma, F.; Tao, Y. Study on Effect of Nanobubble on Ultra-Fine Flake Graphite (UFG) Flotation. *Part. Sci. Technol.* **2023**, *41*, 1062–1070. [CrossRef]
171. Yoon, R.-H. The Role of Hydrodynamic and Surface Forces in Bubble-Particle Interaction. *Int. J. Miner. Process.* **2000**, *58*, 129–143. [CrossRef]
172. Dai, Z.; Fornasiero, D.; Ralston, J. Particle-Bubble Collision Models—A Review. *Adv. Colloid Interface Sci.* **2000**, *85*, 231–256. [CrossRef]
173. Zhou, R.; Wang, H.; Li, X.; Li, D.; Wang, W.; Liang, Y.; Yan, X.; Zhang, H. Effect of Energy Input on Flotation of Particles with Different Sizes: Perspective of Hydrodynamics Characteristics. *J. Environ. Chem. Eng.* **2023**, *11*, 111272. [CrossRef]
174. Ye, J. Study on Interfacial Regulation of Fine Apatite Flotation. Ph.D. Thesis, Guizhou University, Guiyang, China, 2020.
175. Yang, B.; Huang, P.; Song, S.; Luo, H.; Zhang, Y. Hydrophobic Agglomeration of Apatite Fines Induced by Sodium Oleate in Aqueous Solutions. *Results Phys.* **2018**, *9*, 970–977. [CrossRef]
176. Zhao, L.; Zhang, Q. Study of Oleic Acid-Induced Hydrophobic Agglomeration of Apatite Fines through Rheology. *Miner. Eng.* **2024**, *218*, 108911. [CrossRef]
177. Basařová, P.; Hubička, M. The Collision Efficiency of Small Bubbles with Large Particles. *Miner. Eng.* **2014**, *66*, 230–233. [CrossRef]
178. Cheng, G.; Shi, C.; Yan, X.; Zhang, Z.; Xu, H.; Lu, Y. A Study of Bubble-Particle Interactions in a Column Flotation Process. *Physicochem. Probl. Miner. Process.* **2017**, *53*, 17–33. [CrossRef]
179. Ross, V.E. Flotation and Entrainment of Particles during Batch Flotation Tests. *Miner. Eng.* **1990**, *3*, 245–256. [CrossRef]
180. Hoang, D.H.; Hassanzadeh, A.; Peuker, U.A.; Rudolph, M. Impact of Flotation Hydrodynamics on the Optimization of Fine-Grained Carbonaceous Sedimentary Apatite Ore Beneficiation. *Powder Technol.* **2019**, *345*, 223–233. [CrossRef]
181. Waters, K.E.; Hadler, K.; Cilliers, J.J. The Flotation of Fine Particles Using Charged Microbubbles. *Miner. Eng.* **2008**, *21*, 918–923. [CrossRef]
182. Santana, R.C.; Ribeiro, J.A.; Santos, M.A.; Reis, A.S.; Ataíde, C.H.; Barrozo, M.A.S. Flotation of Fine Apatitic Ore Using Microbubbles. *Sep. Purif. Technol.* **2012**, *98*, 402–409. [CrossRef]
183. Li, C.; Wang, L. Improved Froth Zone and Collection Zone Recoveries of Fine Mineral Particles in a Flotation Column with Oscillatory Air Supply. *Sep. Purif. Technol.* **2018**, *193*, 311–316. [CrossRef]
184. Song, S.; Zhang, X.; Yang, B.; Lopez-Mendoza, A. Flotation of Molybdenite Fines as Hydrophobic Agglomerates. *Sep. Purif. Technol.* **2012**, *98*, 451–455. [CrossRef]
185. Santana, R.C.; Farnese, A.C.C.; Fortes, M.C.B.; Ataíde, C.H.; Barrozo, M.A.S. Influence of Particle Size and Reagent Dosage on the Performance of Apatite Flotation. *Sep. Purif. Technol.* **2008**, *64*, 8–15. [CrossRef]

186. Jiang, H.; Yang, H.; Xing, Y.; Cao, Y.; Gui, X. The Effect of Impeller Speeds on the Nanobubbles Flotation Efficiency of Ultrafine Coal Particles. *Powder Technol.* **2025**, *449*, 120431. [CrossRef]
187. Zeng, M.; Li, K.; Huang, L.; Bao, S.; Liu, C.; Yang, S. Interaction Mechanism of Interfacial Nano-Micro Bubbles with Collec-tors and Its Effects on the Fine Apatite Flotation. *Appl. Surf. Sci.* **2025**, *682*, 161736. [CrossRef]
188. Jung, M.U.; Kim, Y.C.; Bournival, G.; Ata, S. Industrial Application of Microbubble Generation Methods for Recovering Fine Particles through Froth Flotation: A Review of the State-of-the-Art and Perspectives. *Adv. Colloid Interface Sci.* **2023**, *322*, 103047. [CrossRef]
189. Chen, Y.; Truong, V.N.T.; Bu, X.; Xie, G. A Review of Effects and Applications of Ultrasound in Mineral Flotation. *Ultrason. Sonochem.* **2020**, *60*, 104739. [CrossRef] [PubMed]
190. Jung, M.U.; Kim, Y.C.; Bournival, G.; Ata, S. An Acoustic Agglomeration Method for Segregation of Micro– to Nano–Bubbles for the Flotation of Ultrafine Particles. *Sep. Purif. Technol.* **2025**, *361*, 131290. [CrossRef]
191. Gao, K.; Liu, H.; Sun, L.; Zhang, Z. Effect of Gas Input Conditions and Ultrasound on the Dynamic Behavior of Flotation Bubbles. *ACS Omega* **2022**, *7*, 22326–22340. [CrossRef] [PubMed]

minerals

MDPI

Article

Recovery of Xenotime and Florencite from Silicate Minerals Using a Combined Technique of Magnetic Separation and Flotation

Weiwei Wang [1,2], Erdou Li [2], Zhengyao Li [1,3,*], Weiyao Zhu [1] and Yuanyuan Wang [4]

1 School of Civil and Resource Engineering, University of Science and Technology Beijing, Beijing 100083, China; viviw91@163.com (W.W.)
2 State Key Laboratory of Baiyun Obo Rare Earth Resource Researches and Comprehensive Utilization, Baotou Research Institute of Rare Earths, Baotou 014030, China
3 State Key Laboratory of High-Efficient Mining and Safety of Metal Mines of Ministry of Education, University of Science and Technology Beijing, Beijing 100083, China
4 Baotou Rare Earth R&D Center, Baotou 014030, China
* Correspondence: zyli0213@ustb.edu.cn

Abstract: Xenotime (YPO4), a significant phosphatic minerl rich in heavy rare earth elements (HREEs), typically associates with granitic rocks, exemplified in the Wolverine rare earth deposit in Australia. A mineral composition analysis indicates that the primary valuable minerals in the deposit are principally xenotime and minor florencite, with quartz and illite as the main gangue minerals, showing a relatively simple mineral composition. The grade of rare earth concentrate was increased to 14.29% and the recovery reached 94.48% through the magnetic separation pre-enrichment test. However, a high-grade rare earth concentrate could not be achieved using magnetic separation alone. Further purification of the magnetic concentrate is conducted through flotation. The grade of rare earth concentrate reached 51.26%, and the recovery rate reached 90.47%. In summary, this process achieves the efficient recovery of xenotime and florencite, having substantial industrial potential.

Keywords: rare earth elements; xenotime; florencite; magnetic separation; flotation

1. Introduction

Rare earth elements (REEs) comprise a group of 17 elements in the IIIB family of the periodic table, encompassing yttrium, scandium, and lanthanides [1,2]. These elements exhibit distinctive spatial structures and physicochemical properties, making them indispensable in various high-tech sectors. Often referred to as the "vitamins of industry", rare earth elements are fundamental raw materials for many key and advanced technological applications [3].

REEs are commonly divided into two groups: light rare earth elements (LREEs) (lanthanum, cerium, praseodymium, neodymium, promethium, samarium, and europium), which possess lower atomic numbers and a smaller mass, and heavy rare earth elements (HREEs) (scandium, yttrium, gadolinium, terbium, dysprosium, holmium, erbium, thulium, ytterbium, and lutetium). Heavy rare earth elements have extensive applications in critical industries, including in catalysts, magnetic materials, colored glass, laser materials, and electronic components, among other fields [4–8]. REEs typically appear as three economical mineral compounds: bastnaesite ($(Ce,La)CO_3F$) and monazite ($(Ce,La,Nd,Th)PO_4$) for LREEs, and xenotime (YPO_4) for HREEs [9]. In this context, the majority of investigations for enriching resources of REEs have concentrated on bastnaesite and monazite, with less focus on xenotime processing [10].

The beneficiation processes for rare earth ores primarily include gravity separation, magnetic separation, and flotation [11–14]. Gravity separation technology is most commonly employed to separate monazite from placer [15]. The magnetic separation process can achieve relatively ideal recovery indicators for rare earth ores with simple properties and a low magnetic mineral content. However, for rare earth ores with complex mineral compositions and fine dissemination sizes, it is difficult to directly obtain high-grade rare earth concentrates through this method alone. Consequently, magnetic separation is often used as a preliminary treatment to pre-concentrate low-grade rare earth ores. Wang et al. [16] explored the feasibility of a strong magnetic pre-concentration for low-grade rare earth ores from Maoniu Ping in Sichuan, achieving a concentration ratio of 4.53 under a magnetic field strength of 1.0 T. Zhang et al. [17] implemented a pre-concentration process consisting of one roughing and one scavenging strong magnetic separation, increasing the grade of rare earth oxides from 0.985% to 2.633%, with a recovery of 85.96%. Strong magnetic pre-concentration technology is one of the potential effective methods for processing rare earth ores.

The flotation process is highly adaptable because it exhibits better recovery efficiency for fine-grained rare earth minerals [18–20]. This technique exploits the differences in hydrophobicity between valuable minerals and gangue to achieve separation. Flotation collectors selectively adsorb onto valuable minerals, enhancing hydrophobic differences [21]. These collectors used for REE mineral flotation are generally classified into three types based on their polar functional groups: fatty acids, phosphoric acids, and hydroxamic acids [22]. Fatty acid collectors are widely used in the flotation of rare earth minerals due to their high collecting performance and availability. However, during the flotation processes of bastnaesite and monazite, these collectors also display a strong affinity to gangue minerals like fluorite, dolomite, and barite, resulting in poor selectivity [18]. Phosphoric acid collectors, on the other hand, react strongly with surface metal ions due to the oxygen atoms in their structure, providing robust collecting capabilities [23]. Hydroxamic acids have been demonstrated as the most effective collectors for the flotation of bastnaesite and monazite and are extensively utilized worldwide [24]. Several previous studies indicate that different structural hydroxamic acid collectors exhibit notable variations in flotation performance for rare earth minerals. For instance, Ren [25] performed experiments on the flotation of bastnaesite using C5~9 alkyl hydroxamic acids and yielded promising separation results. However, these collectors often face challenges related to low selectivity and stability. Cycloalkyl hydroxamic acids, on the other hand, exhibit a promising selectivity for bastnaesites but possess low collecting properties. Among hydroxamic acid collectors, 2-hydroxy-3-naphthylmethyl hydroxamic acid (NHA) is the most widely used in the industrial separation of bastnaesite due to its high efficiency and chemical stability [26]. Despite its advantages, the application of NHA collectors in the flotation of xenotime has not been documented, leaving a gap in the research for optimizing the flotation processes for this specific rare earth mineral.

Most rare earth ores exhibit a heterogeneous distribution of valuable mineral particle sizes, with a significant presence of gangue minerals that are similar in properties to the rare earth minerals. Therefore, employing a single beneficiation process is insufficient for obtaining high-quality rare earth concentrates. Based on the mineral composition, the particle size distribution, and the physical and chemical property differences among various minerals, the rational combination of multiple separation processes such as gravity, magnetic, and flotation techniques is essential. This integrated approach aims to further enhance the comprehensive utilization rate of resources and represents the primary trend in the future development of rare earth beneficiation processes.

This paper proposes an efficient beneficiation method, and systematically studied the influence of the magnetic separation and flotation process parameters and the grinding fineness on the separation and recovery of xenotime and florencite. In addition, the obtained product has been fully characterized. In the current work, X-ray Diffraction (XRD) and the Advanced Mineral Identification and Characterization System (AMICS) are used to analyze the mineral composition of the ore, and the particle size analyzer detects the particle size of the rare earth minerals under different grinding conditions. It is expected to be sustainable and environmentally friendly, and an alternative to the current industrial process.

2. Materials and Methods

2.1. Minerals and Reagents

2.1.1. Minerals

The ore samples containing xenotime and florencite used in this study were from the Wolverine silicate deposit in Australia. The ore samples were first picked up for process mineralogy research, crushed by jaw crushers and roller crushers, screened to obtain mineral particles smaller than 2 mm, and then riffled to produce 5.0 kg batches for grinding, magnetic separation, and flotation tests. One batch was further ground to −74 μm for chemical composition and mineral analysis.

The concentrations of key rare earth elements in the raw materials and magnetic separation–flotation test products were analyzed using Inductively Coupled Plasma Mass Spectrometry (ICP-MS, NexION 300X, PerkinElmer, USA). Other elemental contents were determined using X-ray Fluorescence (XRF). As shown in Table 1, the chemical AZanalysis results of the ore samples indicate that the primary chemical components of the sample are SiO_2 (76.50%), rare earth oxide (REO, 5.94%), P_2O_5 (3.53%), and Al_2O_3 (4.03%). The main rare earth elements Y_2O_3 (3.66%), Dy_2O_3 (0.57%), Gd_2O_3 (0.32%), Er_2O_3 (0.30%), and Yb_2O_3 (0.25%) make up 85.86% of the total rare earth content, indicating the sample is rich in heavy rare earth elements.

Table 1. The main elemental composition of the ore/%.

Element	**SiO_2**	**REO ***	**P_2O_5**	**Al_2O_3**	**Fe_2O_3**	**Y_2O_3**	**La_2O_3**	**CeO_2**	**Pr_6O_{11}**	**Nd_2O_3**
Content	76.50	5.94	3.53	4.03	5.54	3.66	0.07	0.19	0.01	0.13
Element	Sm_2O_3	Eu_2O_3	Gd_2O_3	Tb_4O_7	Dy_2O_3	Ho_2O_3	Er_2O_3	Tm_2O_3	Yb_2O_3	Lu_2O_3
Content	0.14	0.03	0.32	0.08	0.57	0.10	0.30	0.05	0.25	0.04

REO * is total rare earth oxide.

The mineralogical phase analysis of the ore samples was carried out by X-ray Diffraction (XRD), Advanced Mineral Identification, and Characterization System (AMICS), and the results are shown in Figure 1. The primary rare earth mineral is xenotime, constituting 5.26% of the ore. There is also a minor presence of florencite at 0.85%. The predominant gangue minerals are the silicate minerals quartz and illite, with contents of 84.07% and 6.30%, respectively.

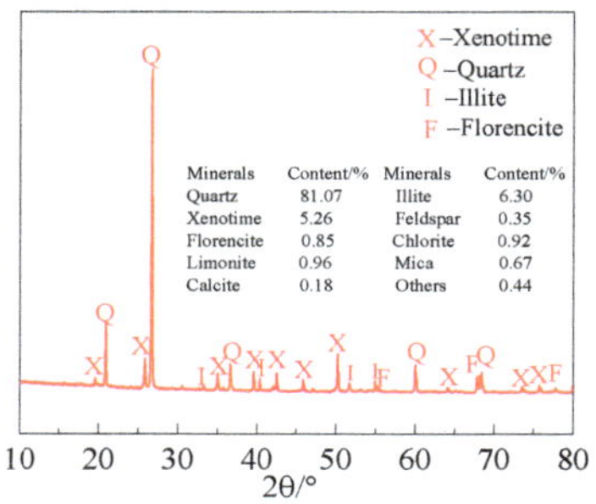

Figure 1. XRD patterns and mineral composition of the ore.

The Scanning Electron Microscope (SEM)—Energy Dispersive Spectrometer (EDS) of raw ore (Figure 2a–f) display the distribution characteristics of xenotime and florencite of rare earth minerals, which are closely associated. Figure 2a shows that xenotime primarily appears as irregular grains embedded in hematite with a particle size range of 3 to 5 μm, while florencite is mainly found in illite, presenting as elliptical or square particles with a diameter of about 5 to 10 μm. Figure 2b shows xenotime embedded in quartz in columnar or needle-like forms with varying grain sizes, while florencite appears in a honeycomb pattern within illite. In Figure 2d, xenotime forms irregular aggregates within quartz fissures, with relatively coarse grain sizes, and these aggregates often contain florencite inclusions. Figure 2e illustrates the coexistence of irregular granular xenotime and fibrous florencite, interspersed with fine-grained quartz particles. The diverse forms and sizes of xenotime (Figure 2f) and florencite (Figure 2c) pose significant challenges in determining the optimal milling grain size for processing.

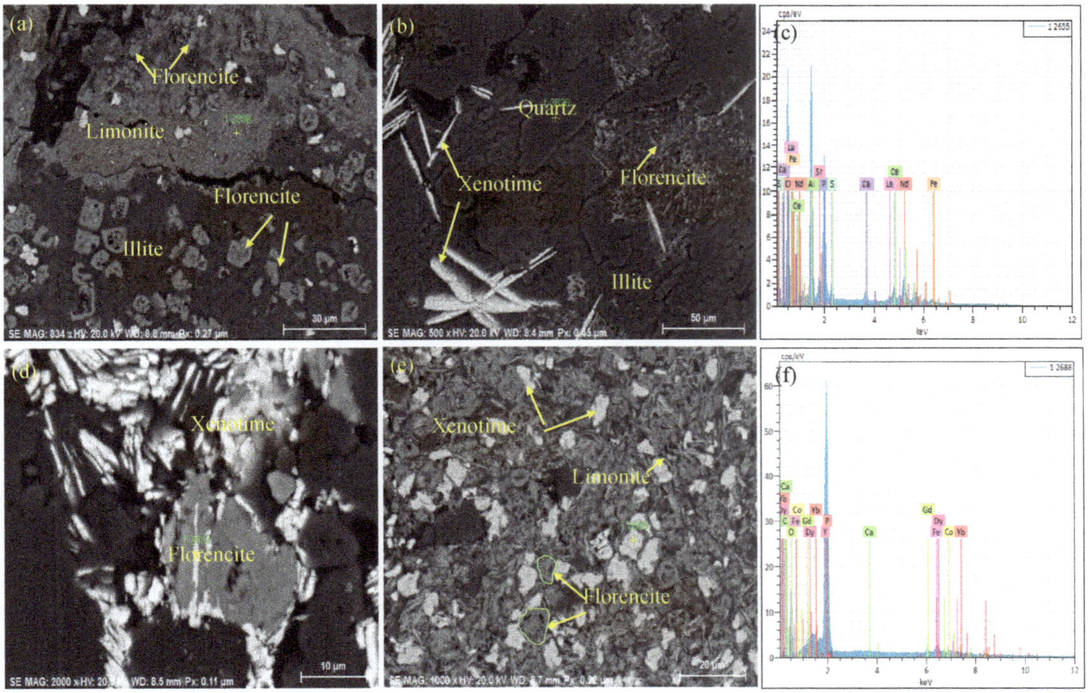

Figure 2. The SEM images of raw ore (**a**,**b**,**d**,**e**), the EDS of florencite (**c**) and xenotime (**f**), respectively.

2.1.2. Reagents

In the flotation tests, sodium oleate (collector NaOL, 98% purity), phenylphosphonic acid (collector SPA, 95% purity), and salicylhydroxamic acid (collector SHA, 95% purity) were sourced from Shanghai Macklin Biochemical Technology Co., Ltd. (Shanghai, China). 2-hydroxy-3-naphthoic acid oxime (NHA collector, 95% purity), sodium silicate (modulus 2.5), and terpineol (foaming agent, 95% purity) were obtained from Baotou Steel Group Baiyun Mineral Processing Branch. Analytical-grade sodium carbonate (Na_2CO_3, Tianjin Kemio Chemical Reagent Co., Ltd., Tianjin, China) were used as pH regulators. All chemical solutions are prepared fresh every day. Tap water in the laboratory was used for the experiments.

2.2. Magnetic Separation Method

In this study, the LHGC-500 high-gradient magnetic separator from Shandong Huatie Magnetic Electric Technology Co., Ltd. (Weifang, China) was employed. When the separator is operational, a large direct current is fed through the excitation coil, creating a high-strength magnetic field in the separation space. This magnetic field generates substantial magnetic forces on the surface of the magnetic medium. By adjusting the middlings vein mechanism to increase the pulse frequency and peak, the weakly magnetic particles are dislodged from the surface of the magnetic medium under the action of fluid forces and pass through the gaps into the tailings bin. In contrast, particles with stronger magnetic properties experience a magnetic force greater than the fluid dynamics, adhere to the surface of the magnetic medium, rotate with the sorting ring, and, ultimately, enter the concentrate bin.

2.3. Bulk Flotation Method

The experiments were carried out using an XFG-type mechanical stirring flotation machine (Jilin Exploration Machinery Factory, Changchun, China). The machine featured a rougher flotation cell with a capacity of 1.5 L and was operated at an impeller speed of 1992 rpm. Each experiment began by mixing 800 g of the raw material with approximately 1.5 L of water in the flotation cell, followed by 3 min of stirring to create a uniform mineral pulp. During each flotation test, the concentrate and tailings were filtered, dried, and weighed. The recovery rate of REO in the concentrate was calculated using Equation (1), which is based on the quality distribution and grade of the products. The average values from three repeated flotation tests were taken as the experimental values, with standard deviations represented by error bars to indicate variability:

$$\varepsilon_{REO} = \frac{M_C \beta_C}{M_0 \beta_0} \times 100\% \tag{1}$$

where M_C is the mass of the rare earth concentrate; β_C is the grade of the rare earth concentrate; M_0 is the mass of the raw ore; and β_0 is the grade of raw ore.

3. Results and Discussion

3.1. Roughing Magnetic Experiments

The particle size significantly impacts the separation of minerals. Therefore, the influence of the grinding fineness (−0.06 mm) on the magnetism of raw ore was first studied. The results are shown in Figure 3a. It can be noted that, as the proportion of −0.06 mm particles gradually increases, the recovery and grade of rare earths in the concentrate also gradually increase. When the proportion of −0.06 mm particles reaches 81.40%, the recovery and grade of rare earths reach 89.17% and 19.88%, respectively. Further increasing the grinding fineness results in a gradual decrease in the recovery and grade of the concentrate. Considering both the beneficiation indicator and grinding cost comprehensively, the optimal particle proportion for the grinding fineness is set at −0.06 mm accounting for 81.40%.

The magnetic field intensity is a critical parameter in magnetic separation. The feed concentration was set at 20% with mineral particles of −0.06 mm constituting 81.4%. The influence of magnetic field intensities on the magnetic separation was studied. The results are shown in Figure 3b. The experimental results indicate that, at a magnetic field intensity of 8000 Gs, the grade of rare earths in the concentrate was 21.05%, but the recovery was below 90%, resulting in a higher content of rare earths in the magnetic tailings. As the magnetic field intensity increased, the recovery of the magnetic concentrate significantly improved. However, when the magnetic field intensity reached 12,000 Gs, changes in the recovery of the magnetic concentrate became less pronounced, and the grade notably decreased. Therefore, the optimal magnetic field intensity is determined to be 12,000 Gs.

Magnetic polymeric media serve as carriers in the high-gradient magnetic separation and significantly influence the effectiveness of magnetic separation. Under the conditions of a magnetic field intensity of 12,000 Gs and a pulse frequency of 15 Hz, the impact of the media type on the magnetic separation was examined. The results are shown in Figure 3c. The experimental findings demonstrate that, as the diameter of the media increased, the recovery of the magnetic concentrate continuously decreased, but the grade gradually improved. After considering both the recovery and grade, a rod-shaped magnetic medium with a diameter of 1.0 mm was selected.

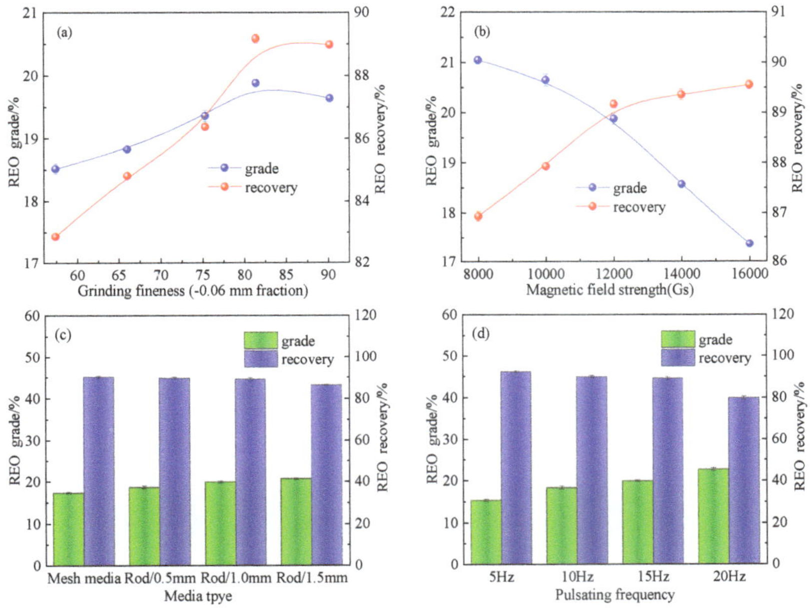

Figure 3. The influence of grinding fineness (**a**), magnetic field intensity (**b**), magnetic medium (**c**), and pulsation frequency (**d**) on the grade and recovery of magnetic concentrate.

Pulsation during the magnetic separation can keep mineral particles in a loose state throughout the sorting process, effectively reducing the mechanical entrapment of gangue particles. Magnetic pulsation experiments were conducted with a grind size of −0.06 mm constituting 81.4%, a magnetic field strength of 12,000 Gs, and 1.0 mm rod-shaped magnetic media. The results are shown in Figure 3d. The experimental outcomes indicate that, with the continuous increase in the pulsation frequency, the grade of rare earths in the magnetic concentrate improved, demonstrating that an increase in pulsation frequency aids in the dispersion of mineral particles and prevents inclusions. However, when the pulse frequency exceeded 15 Hz, the fine minerals adsorbed on the polymagnetic medium may fall off, and the recovery of the rare earth concentrate decreases significantly. Consequently, it was decided that the optimal pulsation for rougher separation is set at 15 Hz.

3.2. Scavenging Magnetic Experiments

The mineral composition and particle size distribution of the roughing magnetic concentrate and tailings were carried out by AMICS and the results are shown in Figure 4. As shown in Figure 4a, the content of xenotime in the tailings was 0.36%, and that of florencite was 0.35%. Figure 4b provides the particle size distribution of xenotime and florencite in the tailings. Florencite has finer particles, with the −20 μm size class accounting for 56.49%, and the 20~45 μm size class accounting for 41.35%. The particle size distribution of xenotime is more uniform, with, notably, 15.22% in the +74 μm size class. To determine the cause of the loss of rare earth minerals, the monomer dissociation characteristics of the rare earth minerals were measured. As can be seen from Figure 4c, the mono-liberation of xenotime reached up to 40.21%. In order to further reduce the content of rare earth in the tailings, a scavenging step was added based on the rougher separation. The experimental process and results are presented in Figure 5. It is evident that, after adding one scavenging step, the rare earth content in the magnetic tailings can be reduced from 0.83% to 0.54%, and the recovery of rare earths in the magnetic concentrate increased from 89.90% to 94.48%.

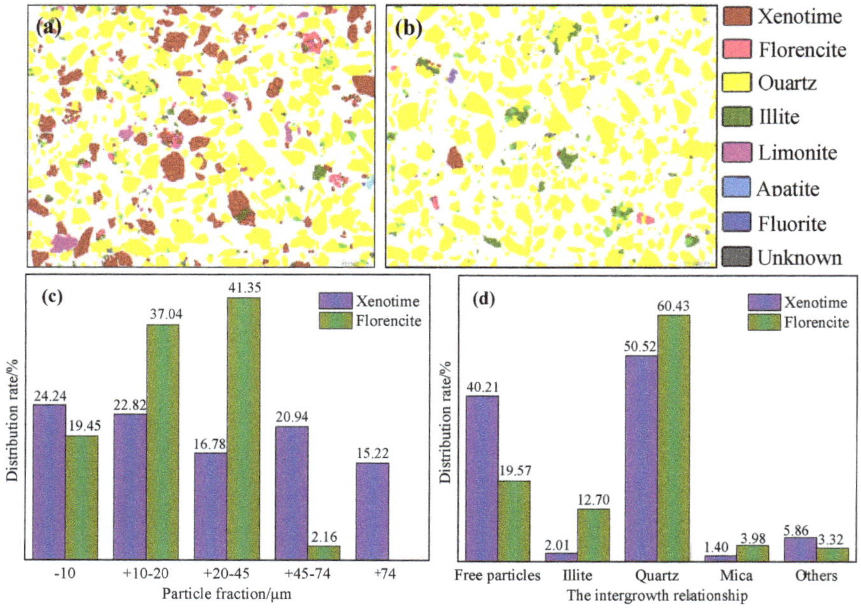

Figure 4. SEM image of roughing magnetic concentrate (**a**), SEM image of roughing magnetic tailings (**b**), the particle size distribution of xenotime and florencite (**c**) in tailings, and the intergrowth relationship of xenotime and florencite (**d**) in tailings.

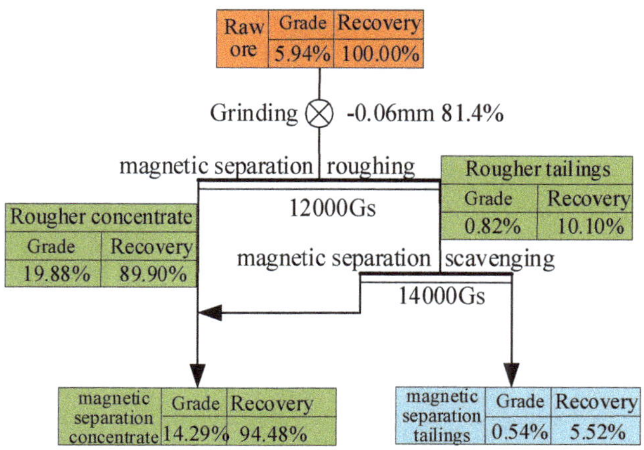

Figure 5. The experimental flowchart of magnetic separation.

3.3. Roughing Flotation Experiments

From the mineralogical research results (Section 2.1.1), it is evident that the rare earth minerals exhibit uneven grain sizes embedded within the ore, making fine grinding crucial for the effective separation of rare earth minerals and gangue minerals. Flotation is the most effective method for processing fine-grained minerals. In the experiments determining the optimal grinding fineness, conditions included 1.8 kg/t of the collector, 0.6 kg/t of sodium silicate, and 0.12 kg/t of pine oil determined in previous research. Figure 6a shows that increasing the grinding fineness significantly improves the grade and recovery of the rare earth concentrate. When the grinding fineness reached 83.56% passing −0.043 mm, the grade of the rough concentrate was 29.36%, with a recovery of 95.04%. However, further increasing the grinding fineness slightly improved the concentrate grade, but significantly decreased the recovery, indicating that the optimal regrinding fineness is 83.56% passing −0.045 mm.

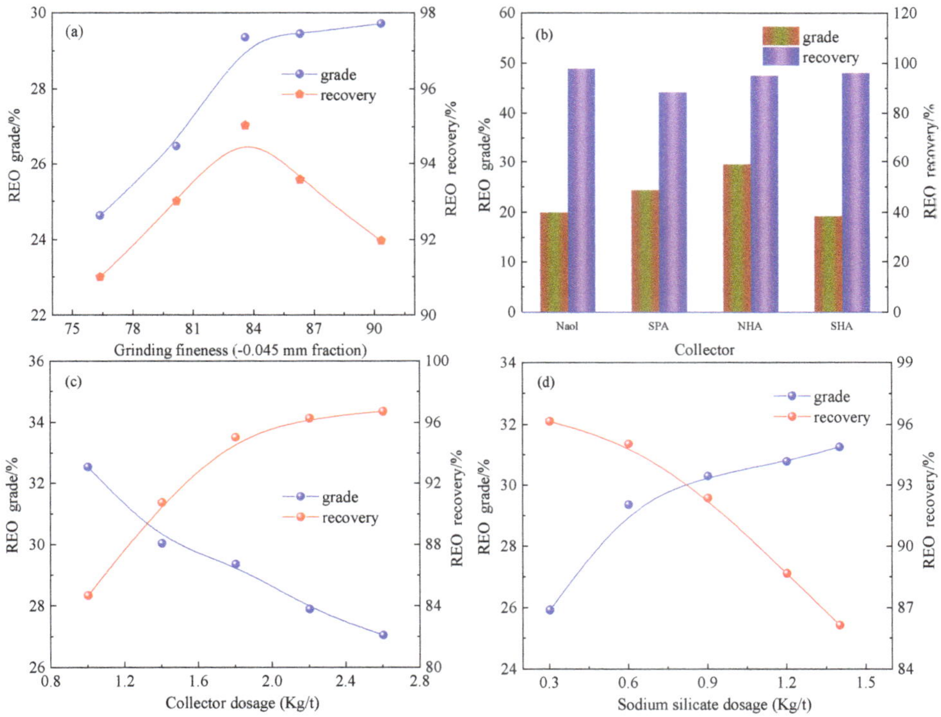

Figure 6. The influence of grinding fineness (**a**), typical reagents (**b**), collector dosage (**c**), and sodium silicate dosage (**d**) on the grade and recovery of flotation rough concentrate.

Figure 6b presents comparative test results of typical reagents such as NaOL, SPA, NHA, and SHA. It is observed that, under the same flotation conditions, NaOL and SHA achieved the highest recovery for the rough rare earth concentrate. However, due to their poor selectivity, the grades of the rough concentrates were only 19.84% and 19.19%, respectively. SPA was found to be inadequate in collecting rare earth minerals. In contrast, using NHA improved both the recovery and grade of the rough rare earth concentrate. The further quantification of the NHA dosage indicated that the recovery of the rough rare earth concentrate initially increased with the increasing amount of the collector (Figure 6c). When the collector dosage reached 1.8 kg/t, the recovery of the rough concentrate peaked at 95.04%. Further increases in the collector dosage did not significantly enhance the recovery but led to a decrease in the grade. Therefore, the optimum collector dosage was determined to be 1.8 kg/t.

Based on a mineralogical analysis, the main gangue minerals identified in the ore sample are quartz and illite, both of which are silicate minerals. Sodium silicate is an effective depressant for these silicate minerals. Figure 6d shows the effect of sodium silicate dosage on the recovery of rare earth elements. When the dosage of sodium silicate was increased from 0.3 kg/t to 0.6 kg/t, the grade of the rough concentrate improved from 25.93% to 29.36%. Further increasing the dosage slightly enhanced the grade, but significantly lowered the recovery. Therefore, the optimum dosage of sodium silicate was determined to be 0.6 kg/t.

3.4. Closed-Circuit Flotation Experiments

On the basis of the roughing flotation tests, closed-circuit flotation validation experiments were conducted using the types and optimal dosages of reagents determined in Section 3.2. The reagents used in the flotation cleaning section were the same as those in the rougher flotation, but at half the dosage. Scavenger flotation used only the collector and frother. The closed-circuit flotation process is shown in Figure 7, and the results of the closed-circuit flotation experiments are presented in Table 2. Table 2 reveals that the obtained rare earth concentrate had a grade of 51.26% with a recovery of 90.47%, while the grade of rare earths in the tailings was only 1.82%. Compared with the results in the literature (Wang et al. [20], and Zhang et al. [21]), the grade and recovery of rare earth concentrates have been significantly improved in the experimental results. The mineral composition analysis of the flotation concentrate is shown in Figure 8. As can be seen from Figure 8, xenotime and florencite are almost exclusively present as free particles, with the mineral contents of xenotime and florencite being 68.26% and 3.39%, respectively. This proves that the effective recovery of rare earth minerals in this ore was achieved through the combined process of magnetic separation and flotation.

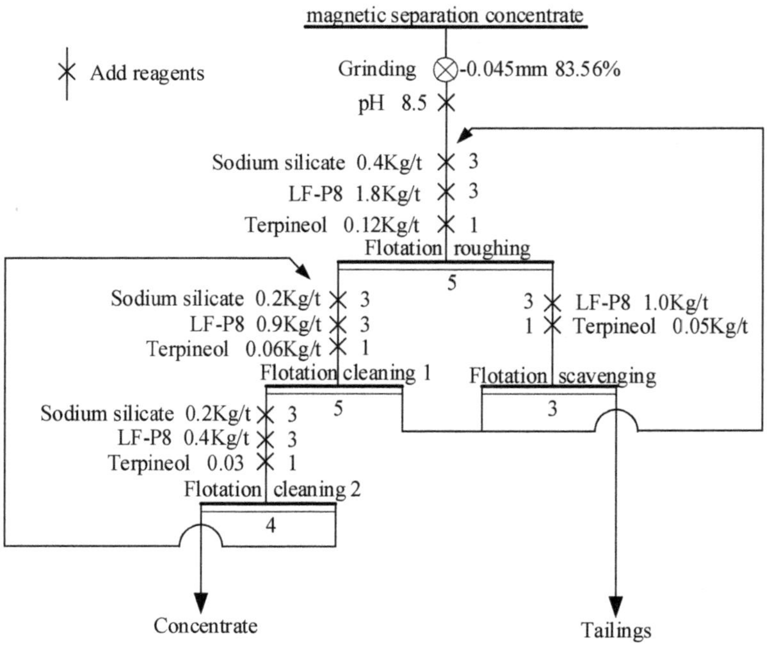

Figure 7. Process flowchart of closed-circuit flotation experiments.

Table 2. Experimental results of comprehensive conditions/%.

Sample	REO		
	Yield	Recovery	Grade
Flotation concentrate	25.22	90.47	51.26
Flotation tailing	14.05	4.01	1.82
Magnetic concentrate	39.27	94.48	14.29
Magnetic tailings	60.73	5.52	0.54
Raw ore	100.00	100.00	5.94

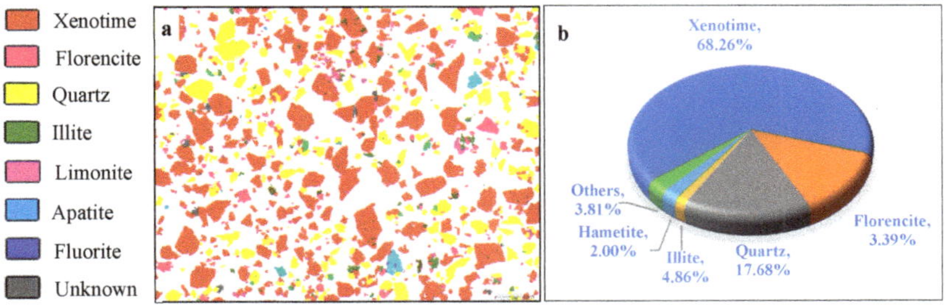

Figure 8. SEM image (**a**) and mineral composition (**b**) of the flotation concentrate.

4. Conclusions

This work proposes an efficient recovery process of rare earth minerals. The grade of REO is 5.94% in the ore, and the rare earth minerals are xenotime and florencite, with contents of 5.26% and 0.85%, respectively. Due to the uneven grain sizes of rare earth minerals, it is necessary for a staged grind and staged separation to avoid the over-pulverization of the ore. The results indicate that, when the grinding fineness of −0.060 mm comprises 81.40%, using only magnetic pre-enrichment can increase the grade of the rare earth concentrate from 5.94% to 14.29%, while reducing the REO content in the tailings to 0.54%. The further purification of the magnetic concentrate is conducted through flotation. The flotation process included one regrind, one rougher stage, two cleaner stages, and one scavenger stage with middlings recycled in a closed circuit, achieving an REO grade of 51.26% with a recovery of 90.49%. Simultaneously, the coexistence relationships and liberation degrees of rare earth minerals in the samples were characterized, and the distribution characteristics of rare earth minerals during magnetic and flotation processes were analyzed.

Author Contributions: Conceptualization, methodology, software, writing—review and editing, funding acqui-sition, Z.L.; resources, formal analysis, investigation, data curation, writing—original draft preparation W.W.; software, E.L.; visualization, W.Z.; supervision, validation, Y.W. All authors have read and agreed to the published version of the manuscript.

Funding: This work was financially supported by the Natural Science Foundation of Inner Mongolia of China (2024MS05061, 2023LHMS05050), and the Program of State Key Laboratory of Bayan Obo Rare Earth Resource Researches and Comprehensive Utilization (2022Z2412).

Data Availability Statement: Not applicable.

Conflicts of Interest: The authors declare no conflict of interest.

References

1. Jung, M.; Tadesse, B.; Dick, C.; Logan, A.; Dyer, L.; Albijanic, B. Influence of monovalent and divalent cations on monazite flotation. *Colloids Surf. A Physicochem. Eng. Asp.* **2022**, *653*, 129975. [CrossRef]
2. Bau, M. Rare-earth element mobility during hydrothermal and metamorphic fluidrock interaction and the significance of the oxidation state of europium. *Chem. Geol.* **1991**, *93*, 219–230. [CrossRef]

3. Wang, W.; Peng, Z.; Guo, C.; Li, Q.; Liu, Y.; Hou, S.; Jin, H. Exploring rare earth mineral recovery through characterization of riebeckite type ore in Bayan obo. *Heliyon* **2023**, *9*, e14060. [CrossRef]
4. Alves, R.C.; Nascimento, M.; Paulino, J.F.; Afonso, J.C. Selection of a hydrometallurgical process for rare earths extraction from a Brazilian monazite. *Hydrometallurgy* **2021**, *200*, 105556. [CrossRef]
5. Cen, P.; Bian, X.; Liu, Z.; Gu, M.; Wu, W.; Li, B. Extraction of rare earths from bastnaesite concentrates: A critical review and perspective for the future. *Miner. Eng.* **2021**, *171*, 107081. [CrossRef]
6. Dürger, N.B.; Anik, M. Synthesis of La_2MgNi_9 hydrogen storage alloy in molten salt. *Int. J. Hydrogen Energy* **2020**, *45*, 8750–8756. [CrossRef]
7. Li, Y.; Zou, D.; Chen, J.; Deng, Y.; Li, D. Stepwise recovery of cerium and fluorine from bastnaesite: Utilizing complex properties of B-F to obtain high purity CeO_2 and KBF_4. *Sep. Purif. Technol.* **2023**, *310*, 123152. [CrossRef]
8. Zhou, H.; Xu, L.; Wang, D.; Xue, K.; Tian, J. Selective adsorption mechanism of SHMP onto fine fluorite in bastnaesite flotation system. *Colloids Surf. A Physicochem. Eng. Asp.* **2023**, *670*, 131527. [CrossRef]
9. Zhu, Z.; Pranolo, Y.; Cheng, C.Y. Separation of uranium and thorium from rare earths for rare earth production—A review. *Miner. Eng.* **2015**, *77*, 185–196. [CrossRef]
10. Marion, C.; Li, R.; Waters, K.E. A review of reagents applied to rare-earth mineral flotation. *Adv. Colloid Interface Sci.* **2020**, *279*, 102142. [CrossRef]
11. Li, W.; Wu, F.; Yang, F.; Lu, W. New research progress in mineral-type rare earth mineral separation process and flotation collector mechanism in China. *Mod. Min.* **2019**, *35*, 1–5.
12. Ma, Y.; Li, N.; Wang, Q.; Yang, Q. Characteristics and current research situation of rare earth resources in Bayan Obo ore. *J. Chin. Soc. Rare Earths* **2016**, *34*, 641–649.
13. Wang, X.; Lin, H.; Dong, Y.; Zhang, Y.; Zhou, S.; Xu, X. Recovery of rare-earth, iron, niobium and fluorite in rare-earth tailings with different magnetic separation and flotation technologies. *Chin. J. Rare Met.* **2014**, *38*, 846–854.
14. Wang, X.; Wang, Y.; Yu, C.; Kang, J. Research progress on mineral processing technology and flotation collector of rare earth. *China Min. Eng.* **2020**, *49*, 73–76.
15. Gupta, C.K.; Krishnamurthy, N. Extractive metallurgy of rare earths. *Int. Mater. Rev.* **1992**, *37*, 197–248. [CrossRef]
16. Cheng, W.; Zhen, H.; Xian, Q.; Tai, W.; Han, L. Practicability research on pre-concentration process by HIMS for bastnaesite type rare earth ore. *Chin. Rare Earths* **2016**, *37*, 56–62.
17. Xin, Z.; Chun, L.; Zhi, L.; Li, L.; Ming, Z.; Ying, M. Pre-concentration of complicated rare-earth mineral bearing ore by wet electromagnetic separation and superconducting magnetic separation. *At. Energy Sci. Technol.* **2022**, *56*, 225–234.
18. Jordens, A.; Cheng, Y.P.; Waters, K.E. A review of the beneficiation of rare earth element bearing minerals. *Miner. Eng.* **2013**, *41*, 97–114. [CrossRef]
19. Dutta, T.; Kim, K.H.; Uchimiya, M.; Kwon, E.E.; Jeon, B.H.; Deep, A.; Yun, S.T. Global demand for rare earth resources and strategies for green mining. *Environ. Res.* **2016**, *150*, 182–190. [CrossRef]
20. Wang, Z.; Wu, H.; Yang, J.; Tang, Z.; Luo, L.; Shu, K.; Xu, Y.; Xu, L. Selective flotation separation of bastnaesite from calcite using xanthan gum as a depressant. *Appl. Surf. Sci.* **2020**, *512*, 145714. [CrossRef]
21. Yang, Z.; Bian, X.; Wu, W. Flotation performance and adsorption mechanism of styrene phosphonic acid as a collector to synthetic $(Ce, La)_2O_3$. *J. Rare Earths.* **2017**, *35*, 621–628. [CrossRef]
22. Yang, X.; Satur, J.V.; Sanematsu, K.; Laukkanen, J.; Saastamoinen, T. Beneficiation studies of a complex REE ore. *Miner. Eng.* **2015**, *71*, 55–64. [CrossRef]
23. Bahsaine, K.; Mekhzoum, M.E.M.; Benzeid, H.; Qaiss, A.E.K.; Bouhfid, R. Recent progress in heavy metals extraction from phosphoric acid: A short review. *J. Indust Eng. Chem.* **2022**, *115*, 120–134. [CrossRef]
24. Liu, W.; Wang, X.; Miller, J.D. Collector Chemistry for Bastnaesite Flotation-Recent Developments. Miner. Process. *Extr. Metall. Rev.* **2019**, *40*, 370–379.
25. Jun, R.; Shouci, L. Mechanism of flotation of bastnaesite by 1-hydroxy-2-naphthalene hydroxamic acid. *Chin. J. Nonferrous Met.* **1996**, *4*, 24–32.
26. Jun, R.; Shouci, L. Study on behavior N-hydroxyl phat halicimide and Napht hyl hydroxamic acid in bastnaesite flotation. *Min. Metall.* **1997**, *6*, 4.

minerals

MDPI

Article

Recovery of Copper from Slags Through Flotation at the Hernán Videla Lira Smelter

Luis Valderrama [1,*], Jaime Tapia [2], Osvaldo Pavez [1], Mario Santander [1], Víctor Rivera [3] and Miguel Gonzalez [4]

1 Department of Engineering in Metallurgy, University of Atacama, Copiapó 1532297, Chile; osvaldo.pavez@uda.cl (O.P.); mario.santander@uda.cl (M.S.)
2 Faculty of Engineering and Architecture, University Arturo Prat, Iquique 1110939, Chile; jtapia@unap.cl
3 National Mining Company (ENAMI), Copiapó 2536199, Chile; vrivera@enami.cl
4 Planta Cerrillos del Grupo Minero Carola Coemin, Vitacura 7630413, Chile; miguel.gonzalez@coeminsa.cl
* Correspondence: luis.valderrama@uda.cl

Abstract: The significant volume of copper smelting slags poses environmental challenges, particularly concerning soil and surface water contamination. However, these slags contain valuable elements such as copper and iron, the recovery of which can contribute to both environmental protection and the circular economy in Chile. This study analyzes, at both laboratory and industrial scales, the recovery of copper sulfides from the slags of the Hernán Videla Lira Smelter in Atacama, Chile. Physical, chemical, and mineralogical characterizations were performed, along with flotation tests (rougher and cleaner) to optimize the grinding degree, pH, reagents, and flotation times. The slag, with a copper grade of 0.71%, contains fayalite, magnetite, quartz, pyrite, and chalcopyrite. In the laboratory, the concentrate obtained in the rougher stage showed a copper grade of 3.7% with a recovery rate of 62.1%; in the cleaner stage, the grade increased to 24.4%, with a recovery rate of 71.7%. At the industrial level, the rougher–cleaner circuit produced concentrates with a copper grade of 27.9% and a recovery rate of 87.5% processing 1344 tons per day, thus demonstrating the viability of this methodology.

Keywords: copper slag; flotation; recovery

1. Introduction

Copper slag is a massive metallurgical residue obtained from transforming copper ore concentrates into anodic copper in smelters. Slags are deposited in landfills that occupy large areas of land. The main environmental impacts of slag disposal are the change in land use and visual contamination of the landscape. On the other hand, under certain climatic conditions, leaching can occur, depending on the characteristics of the solution, the composition, and the final crystalline structure of the solid slag.

Copper slags are substantial metallurgical wastes generated through pyrometallurgical processes and deposited in landfills. Their chemical composition is rich in iron, silicon, and aluminum oxides. Worldwide, slag production in the copper industry is approximately 37.7 million tons, of which 4.5 million tons are attributed to Chilean producers [1,2]. The recovery of metals from copper slags and the utilization of these slags are crucial not only for saving metal resources but also to protect the environment [3].

The flotation of copper slags presents an attractive alternative, considering the copper grades of these wastes, which for the most part, are equal to or higher than the copper ore grades and have been decreasing as the deposits are exploited. In this context, it should be noted that several methods have been documented for the recovery of copper from slags, including physical separation methods, such as flotation and magnetic separation, chemical separation processes, such as reduction, roasting, leaching, and smelting, and combined chemical and physical methods such as reduction and magnetic separation [4].

In the copper slag flotation process, one of the critical factors influencing grinding and flotation is slag cooling after it has been produced in the smelter and deposited in the dump. In this regard, several researchers have reported that slow cooling of the slag can improve copper recovery [5–7]. This technological procedure yields better metallurgical results because it promotes the growth or coagulation of dispersed sulfide and metallic copper particles in the slag, increasing the recovery of the flotation process with a decrease in copper losses through excessively fine particles [8].

Numerous studies on copper slag flotation have been documented in the literature. Refs. [9,10] studied the recovery of copper contained in slag by flotation using sodium isopropyl xanthate as a collector and a particle size of less than 100 μm, obtaining a concentrate with a copper grade of 21% and a recovery rate of 80%. Ref. [11] studied the effect of milling time on copper recovery, finding that 65 min of milling (d80 = 70 μm) resulted in a copper recovery rate of 62.23%, whereas extending the milling time to 85 min (d80 = 48 μm) increased the copper recovery rate to 79.89%. Additionally, the use of collectors individually, as in mixtures of two collectors, was reported by [12], using sodium isopropyl xanthate, sodium diethyl dithiophosphate, and alkyl hydroxamate as collectors in various dosages for copper slag flotation, obtaining the best results with the combination of these reagents. The researchers achieved a higher copper recovery rate of 84.82% using a 40:160 gt^{-1} mixture of sodium isopropyl xanthate with diethyl dithiophosphate, compared to 78.11% obtained with the best single collector. An improvement in recovery reaching 83.07% was also obtained using a 160:40 gt^{-1} mixture of sodium isopropyl xanthate with alkyl hydroxamate. Additionally, the synergistic effect of collectors on copper recovery from slag was also reported by [13], where it was stated that using 10 gt^{-1} Z6 as a collector and 50 gt^{-1} Aero 3477 as a depressant resulted in a copper recovery rate of 62.23% under optimal conditions. In contrast, under optimal conditions using a mixture of 6 gt^{-1} Z6, 4 gt^{-1} Z11, 30 gt^{-1} Aero 3477, and 20 gt^{-1} Aero 2 achieved a copper recovery rate of 80.27%, which indicated a positive synergistic effect of the collector mixture. Ref. [14] studied a range of commercial reagents for copper slag flotation, such as xanthates, dithiophosphates, mercaptobenzothiazole, thionocarbamates, fatty acids, sulfides, and sulfates, indicating that the best flotation performance was achieved with the collector reagent FC 4146, which contains thionocarbamate and methyl isobutyl carbinol as a frother. Ref. [15] reported that, in copper slag flotation, sodium alkyl dithiophosphate and mercaptobenzothiazole (FC7245) were the secondary flotation reagents that yielded the best copper recoveries. Additionally, ref. [16] studied the addition of 600 gt^{-1} of sodium carbonate prior to grinding the copper slag and conditioning with 300 gt^{-1} of xanthate, demonstrating that the copper recovery can reach 98.07 % with a concentrate grade of 36.13%. They indicated that the addition of sodium carbonate increases pulp stability and improves flotation kinetics.

The scientific justification of the flotation process applied in this study is based on two aspects: (a) the slow scheme applied in the cooling of the slag in the foundry; and (b) the chemical reagents used (collectors). The scheme consisting of slow cooling–crushing–grinding–flotation is recommended for the objective of obtaining a high-grade copper concentrate and achieving an efficient flotation process. Therefore, the correct application of a slag cooling method is important. In the slow cooling applied to the slag in this study, the growth or coagulation of dispersed copper and copper sulfide particles in the slag was favored, thus increasing the efficiency of the flotation process, with a decrease in copper losses through very fine particles. The chemical reagents applied in the flotation process, depended on the characteristics of the copper slag, anionic collectors of the thio-compound type (xanthates, dithiophosphates, mercaptobenzothiazole, thionocarbamates) or the non-ionizable thio-compound collectors (fatty acids, sulphates, hydroxamate).

In recent years, at the Manuel Antonio Matta Plant (MAMP) operated by Empresa Nacional de Minería (ENAMI), the treatment of copper sulfide ores has decreased significantly, from 100,000 t to 40,000 t in 2017, which resulted in a decrease in the amount of copper concentrate produced, and an increase in the cost of production. Due to the ore supply issues generated at the concentrator plant, MAMP decided to float copper slag from the Hernán Videla Lira Smelter (HVLS), also owned by ENAMI. This new strategy was adopted to increase the inventory of copper concentrates and included an evaluation and implementation of the flotation of copper slag from the HVLS at MAMP using an existing flotation circuit at the plant for the treatment of these metallurgical wastes.

This paper describes the evaluation of the copper slag flotation conducted at the MAMP concentrator plant, which was developed from flotation tests at the laboratory scale and taken to the industrial scale, and was achieved with the application of a new flotation circuit. All these modifications, made at the MAMP concentrator plant for the treatment of slag from the HVLS, were designed to achieve concentrates with a minimum copper grade of 25%.

2. Materials and Methods

2.1. Material

In this study, a copper slag sample was used from the reverberatory furnace of the HVLS collected from the slag deposit. The approximately 30-ton sample was subjected to successive stages of size reduction to 100% < 1.8 mm.

The slag sample was characterized by determining the specific gravity, work index (Wi), abrasion index (Ai), and chemical and mineralogical analyses. The specific gravity was determined using the pycnometer method. The work index was determined by applying the standard Bond test in a 305 mm × 305 mm (D × L) ball mill, rotating at 70 rpm. The abrasion index was also determined using the method proposed by Bond in a 305 mm × 114 mm (D × L) ESSA abrasion mill, measuring the mass loss of an SAE 4325 steel plate with dimensions equal to 76.2 mm × 25.4 mm × 6.35 mm (L × W × T).

Chemical characterization was performed using volumetric, gravimetric, and GBC 908 atomic absorption spectrophotometer (AAS) methods. The quantitative mineralogical composition was performed by X-ray diffraction (XRD) in a Shimadzu XRD6100 diffractometer and QEMSCAN automated scanning electron microscopy.

2.2. Laboratory-Scale Flotation Tests

Flotation studies were conducted a the laboratory scale at the Department of Metallurgical Engineering, University of Atacama to determine the feed particle size effect on the copper grade and recovery in the concentrate in the rougher flotation stage. Subsequently, a kinetic study was conducted to determine the optimal flotation time for the rougher, cleaner, and scavenger stages.

Prior to the flotation studies, grinding tests were carried out to determine the time required to achieve particle sizes of 60%, 65%, 70%, 75%, and 80% < 75 micrometers. The grinding tests were carried out using a 175 mm × 235 mm (D × L) stainless-steel ball mill, applying a rotation speed of 74 rpm and a solids percentage of 70%. The mill was loaded using 940 g of slag and 10.000 g of 25.4 mm-diameter balls.

Rougher flotation tests were carried out using a Denver D-12 flotation machine with a capacity of 2.5 L. After grinding, the suspension was transferred to the flotation cell, and sufficient water was added to achieve a solids percentage of 35% and then agitated at an impeller rotation speed of 1300 rpm. The pulp pH was adjusted to 8, using lime, to depress the pyrite and aluminosilicates. Subsequently, 40 gt^{-1} of Hostaflot E-703 as a collector and

30 gt^{-1} of Flotanol H-70 as a frother were added to the pulp and conditioned for 5 min. Following the conditioning process, the pulp was floated for 15 min with an air flow of 7 $L{\cdot}min^{-1}$. In copper slag flotation, the consumption of reagents (collector and frother) is increased compared to the dosages used in copper ore concentrator plants [17].

The rougher flotation kinetics were performed under the same experimental conditions as conventional rougher flotation tests. The pulp was floated for 15 min and the mineralized froths were recovered by extracting the froth layer every 15 s at the following time intervals: 0–1; 1–2; 2–3; 3–5; 5–7; 7–10; and 15 min.

The procedure for the scavenger and cleaner flotation kinetics was the same as that employed in the rougher kinetics, except that no flotation reagents were added. Mineralized froths were removed from the froth layer in the scavenger and cleaner kinetics at the following time intervals: 0–1; 1–3; 3–5; and 5–10.

2.3. Industrial Scale Operation

The flow diagram used at the Manuel Antonio Matta Plant to treat slags in the first industrial-level campaigns was as follows. The plant operated with three mills, two Marcy ball mills measuring 2.7 × 2.7 meters, with discharge through grates, and a Pegson Marcy ball mill measuring 2.6 × 2.7 meters with discharge through grates that worked in a closed circuit with hydrocyclones. The overflow fed a pulp-conditioning tank, where the collector reagents and a frother for flotation were added. The flotation circuit used 8 Dorr Oliver cells of 8 m^3 for the rougher stage, and a cleaner circuit consisting of 2 Dorr Oliver cells of 8 m^3 and 2 for the scavenger stage. The copper concentrate obtained was thickened and filtered in a ceramic filter with 10 discs, each of which was 30 centimeters in diameter.

The industrial-scale MAMP plant was modified based on the results obtained from the initial campaigns and is shown in Figure 1. It consists of a grinding and classification circuit, which includes a Hardinge ball mill measuring 3.8 × 5.5 meters and a battery of five hydrocyclones of 50 centimeters each.

The flotation circuit comprises a 12 m^3 conditioner, seven 16 m^3 cells, and one 40 m^3 cell in the rougher stage. One column cell of 2.5 m diameter and a height of 12 m in the cleaner stage, and four cells of 8 m^3, six cells of 3 m^3 and one cell of 20 m^3 in the scavenger stage. Additionally, a 2.1 m × 2.1 m mill operating in a closed circuit with a battery of two hydrocyclones of 25 centimeters used for the regrinding of both rougher and scavenger concentrates.

The operating conditions at the industrial scale were as follows: a feed flow of 70 t/h; a grinding degree of 70% < 74 micrometers; solids by weight feed to flotation of 30%; a pH of 8, a Hostaflot E-703 as a collector (modified thionocarbamate) with a dosage of 40 gt^{-1}; and a Flotanol H-70 (mixture of polyglycolic alcohols and ethers) as a frother, with a dosage of 30 g/t [18].

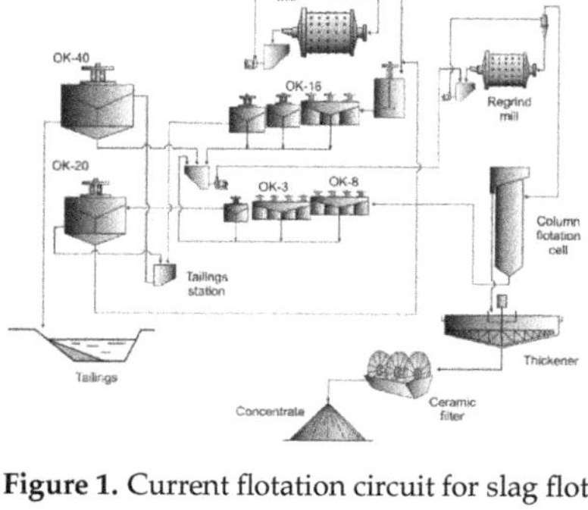

Figure 1. Current flotation circuit for slag flotation.

3. Results

3.1. Physical Characterization

Table 1 presents the physical characterization of the copper slag from the HVLS. The table shows that the actual density, work index, and abrasion index reached values of 3.3 g·mL^{-1}, 22.6 ± 1.1 KWh·t^{-1}, and 0.912 g·h^{-1}, respectively. Generally, copper slags are characterized by densities higher than 3.0 g·mL^{-1} due to their main constituent compounds. For instance, slag samples from the same smelter reported that the actual density of the slag was 3.8 kg·L^{-1}. The high abrasion index of copper slag (0.912 g·h^{-1}) contributes to high grinding media consumption of the ball mills of the plant, compared to the indexes for copper, molybdenum, lead, silver, zinc, gold, and iron ores, where the values fluctuate between 0.08 and 0.70 [19,20].

Table 1. Physical characterization of copper slag.

Density (kg·L^{-1})	Work Index (kWh·t^{-1})	Abrasion Index (g·h^{-1})
3.3	22.6 ± 1.1	0.912

The high steel consumption achieved in the treatment of copper slag is due to the high Bond index. This issue is mainly because of the fast, or slow, controlled cooling effect [21,22].

The chemical analysis results of the slag, presented in Table 2, indicate a copper grade of 0.71 % and the presence of iron and silicon as the most relevant components, in addition to calcium, aluminum, sodium, magnesium, potassium, and sulfur. These findings are corroborated by the diffractogram shown in Figure 2, which shows that the main typical crystalline phases of copper slag are magnetite, pyrite, quartz, and chalcopyrite.

Table 2. Chemical composition of slag.

Component	Cu	Fe	S	SiO_2	CaO	MgO	Al_2O_3	Na_2O	K_2O
Content, %	0.71	34.8	0.53	39.5	8.1	1.5	7.8	2.2	1.3

Figure 2 shows the diffractogram of the slag, in which it can be seen that these slags are made up of both crystalline and vitreous phases; chalcopyrite, clinochlore, cuprospinel, fluorapatite; magnesio-hornblenda, magnetite, pyrite, quartz, and talc are present.

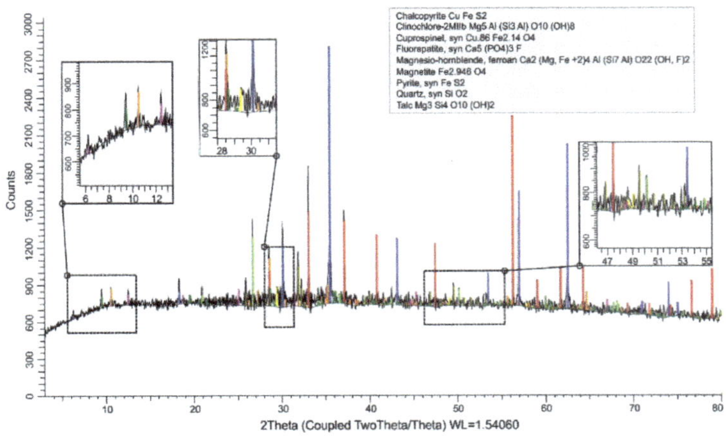

Figure 2. X-ray diffraction of the reverberatory slag.

Conversely, the QUEMSCAN analysis confirms that the slag contains 50.3% magnetite, 22.2% pyrite, 9.7% fayalite, 5.8% hornblende, 4.8% clinochlore, 4.7% quartz, and 1.1% chalcopyrite.

3.2. Flotation Tests

The results of the laboratory flotation tests are shown in Figure 3, where the particle size effect on the copper grade and recovery in the concentrate can be observed. These tests indicate that an increase in slag fineness does not guarantee a significant effect on the recovery or grade of the concentrate obtained, where recoveries did not exceed 60% with copper grades in concentrates below 5%; as a consequence, a cleaning stage is necessary to achieve a better quality of the concentrate. On the other hand, the best result was achieved with a particle size of 70% less than 75 micrometers, a recovery rate of 62.2%, and a copper grade in the concentrate of 3.7% [20]. This low recovery could be due to the low feed grade and the chemical state of copper present in this material. The figure also indicates that a lower granulometry does not necessarily result in higher recovery. This may be attributed to the greater amount of slimes; these ultrafine particles could produce a coating of hydrophobic particles, decreasing their buoyancy [23].

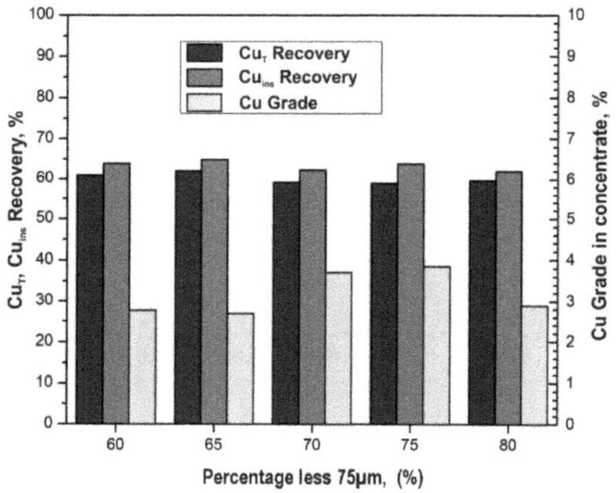

Figure 3. Copper grade and recovery as a function of particle size.

3.3. Flotation Kinetics

Considering the results of the rougher flotation tests regarding the effect of particle size, flotation kinetics was performed for the rougher and cleaner stages to determine the optimum time to apply the criteria of [24]. The optimum flotation time was 7 min, with a copper grade of 3.7% and a recovery rate of 62.1%; likewise, the cleaner flotation kinetics indicated a flotation time of 5 min, and the copper grade was 24.4%, with a recovery rate of 71.7%, such results were published in [20]. Such experimental conditions allow deducing that it is required to include a scavenger stage to increase the overall recovery [25]. Based on the operational conditions at the laboratory level and the optimum flotation times of each stage, a conventional flotation circuit was proposed to process the slag.

3.4. Copper Grades and Recoveries Obtained in the First Campaign

Table 3 presents the copper grades and recoveries obtained from the initial flotation circuit during a campaign period of 17 days. It can be observed that the copper grade is low (less than 20%) with an acceptable recovery, which varied between approximately 70 to 80%. During this campaign, 10,587 tons of slag were processed, and 301.6 tons of concentrate were obtained with a copper grade of 16.9% with a standard deviation (SD) of

1.59, and the average copper recovery rate of 73.9% with an SD of 4.63. It also shows that the tailings grade was 0.17% copper with an SD of 0.031.

The results presented in Table 3 show that the copper grade in the final concentrate was not commercially satisfactory, as it ranged between 14.69 and 19.97%, a considerably low value for the concentrate at the smelter entrance. The main limitation of the circuit was the lack of a cleaning stage with which to increase the copper grade in the final concentrate.

The mineralogical analysis of the concentrate indicated that it contains chalcopyrite (42.6%), magnetite (24.4%), pyrite (12.5%), chalcocite (4.6%), fayalite (5.8%), quartz (3.2%), hornblende (3.7%), and clinochlore (2.4%).

Table 3. Copper grades and recoveries obtained in the slag flotation in the first campaign.

Item/Day	**1**	**2**	**3**	**4**	**5**	**6**	**7**	**8**	**9**
Grade Cu, F.	0.65	0.65	0.63	0.66	0.66	0.64	0.66	0.66	0.67
Grade Cu, C.	15.0	15.0	14.7	17.1	17.1	19.9	15.5	16.5	18.5
Grade Cu, T.	0.20	0.20	0.16	0.15	0.15	0.16	0.20	0.18	0.15
Rec. Cu	69.9	69.9	75.4	78.0	77.9	75.5	70.8	73.7	78.2
Tons, F.	424	538	66	318	616	767	709	723	652
Tons, C.	12.8	16.2	2.1	9.5	18.4	18.5	21.5	21.4	18.5
Item/Day	**10**	**11**	**12**	**13**	**14**	**15**	**16**	**17**	**Total**
Grade Cu, F.	0.67	0.65	0.62	0.67	0.65	0.62	0.67	0.64	0.65
Grade Cu, C.	18.9	16.2	16.7	15.6	19.2	16.3	17.8	15.5	16.9
Grade Cu, T.	0.18	0.13	0.12	0.23	0.19	0.17	0.14	0.22	0.17
Rec. Cu	73.7	80.7	81.1	66.6	71.6	73.4	79.7	66.3	73.9
Tons, F.	627	577	479	686	825	869	863	851	10.587
Tons, C.	16.3	18.8	14.4	19.6	20.2	24.4	25.8	23.2	301.6

Note: F: feed; C: concentrate; T: tailings

Figure 4 shows the copper grades and recoveries achieved in the current flotation circuit (Figure 1), where the cleaner stage and regrinding of the rougher concentrate included a column to improve liberation and achieve a higher quality in the final product for a 22-day campaign. During this time, the plant processed 34.045 tons, with copper grades in the concentrate averaging 25.42% and an average recovery rate of 75.8%. The figure shows superior performance in terms of processing capacity, copper concentrate grade, and copper recovery. There is also an additional improvement in copper recovery by adding a scavenger stage for the rougher tailings. At the PMAM concentrator plant, the flotation circuits do not include a scavenger stage for rougher tailings treatment, owing to the high copper recovery when processing copper sulfide ores.

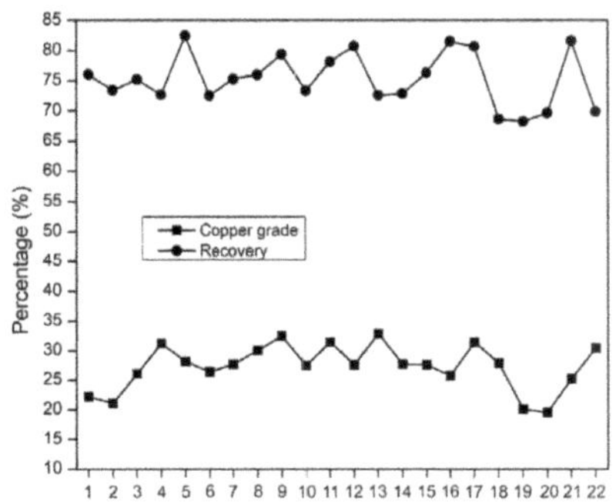

Figure 4. Copper grades and recoveries obtained in the slag flotation in the second campaign.

The higher performance in terms of copper grade is attributed to the columnar cleaning and the rougher concentrate regrinding stages. The improvements achieved in copper recovery by making modifications in the flotation circuit, including a regrinding stage, are reported in [14]. This study showed that the slags were floated in a first stage, obtaining a copper recovery rate of 75.8%, and that in an attempt to improve this recovery, a regrinding stage of the rougher stage tailings was incorporated into the flotation circuit to increase the release of the useful species. Thus, the tailings were reground and floated again, obtaining a recovery rate of 92.2%, representing a significant improvement in copper recovery.

Figure 5 shows the results of slag flotation at MAMP during the third campaign. In February, the concentrate grade has a significant increase, because the feed grade is 4.16%; however, an acceptable recovery rate of 80.7% is maintained. A higher steady increase in copper recovery is observed, with an increase of about 10% compared to the tests in the previous figures.

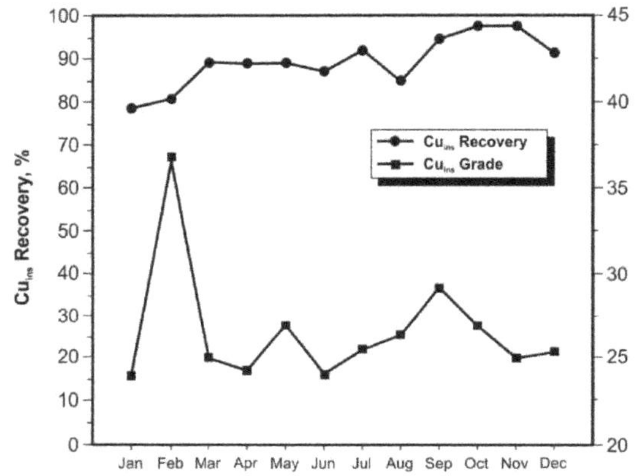

Figure 5. Copper grades and recoveries obtained in the slag flotation in the third campaign.

These results clearly demonstrate that when the slag is initially floated to a size 70% smaller than 74 micrometers in the rougher stage, and this concentrate is reground to a particle size 80% smaller than 44 micrometers, there is a higher liberation of useful species. This increased liberation enhances the selectivity of the flotation process [25].

The results obtained in the slag flotation at MAMP during 2022 are shown in Figure 6. It can be seen that the copper grade and recovery in the concentrate are similar to those of the third campaign.

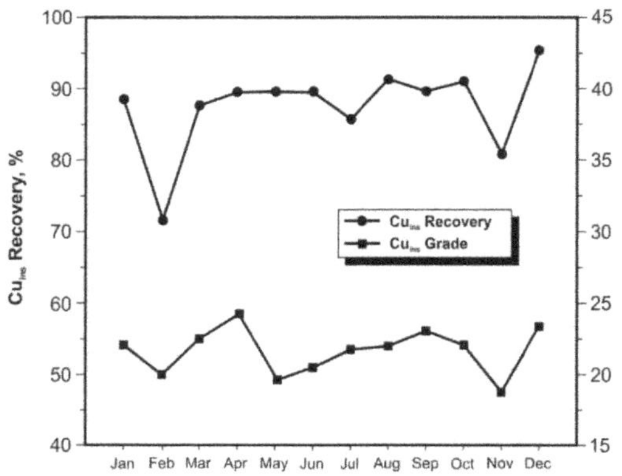

Figure 6. Copper grades and recoveries obtained in slag flotation today.

Table 4 presents a summary of the main parameters of the circuits studied. It can be seen from the data provided in the table that only in the second and third campaigns was the obtained concentrate obtained commercially acceptable, with a copper grade above 25%. On the other hand, from the second campaign onwards, the slag treatment capacity (t/day) was notoriously improved, reaching values approximately two to three times greater than that of the first campaign. As for the metallurgical recovery rate of copper, the values obtained in the first and second campaigns were very similar. However, during the third campaign, and in current operations, recovery increased markedly, by approximately 15 percentage points.

Table 4. Summary of copper grades, capacities, and copper recoveries at MAMP.

Circuit	Copper Grade (% Cu)			Capacity (t/day)	Recovery (%)
	F	C	T		
1st campaign	0.68 ± 0.05	17.36 ± 5.94	0.19 ± 0.03	623.9 ± 198.5	72.93 ± 4.18
2nd campaign	0.93 ± 0.16	24.67 ± 5.29	0.25 ± 0.03	1764.6 ± 655.7	72.90 ± 7.54
3rd campaign	0.86 ± 0.18	27.92 ± 2.59	0.13 ± 0.02	1344 ± 350.3	87.53 ± 5.66
Today	0.65 ± 0.13	21.51 ± 2.75	0.09 ± 0.02	1915 ± 618.4	88.15 ± 7.95

Note: F: feed; C: concentrate; T: tailings.

4. Conclusions

The FHVL slag contained 0.71% copper, which was found in the form of chalco-pyrite, chalcosite, and metallic copper, with a Bond index of 22.6 kWh/t and an abrasion index of 0.9119 g/h, values well above the normal values for this type of copper slag. Rougher flotation tests determined that the optimum particle size distribution was 70% under 75 microns, at a pH of 8, and a dose of 40 g/t of the collector (F-703) and 30 g/t of the frother (H-70). Rougher flotation kinetics indicated that the optimum flotation time was 6 min, with a copper grade of 3.7% copper and a recovery rate of 62.1%. For the cleaner flotation kinetics, the flotation time was 5 min to obtain a copper grade of 24.4% with a recovery rate of 71.7%. These experimental results indicate that a scavenger stage must be included to increase recovery. With the data obtained from the flotation time, open cycle tests were carried out; later, through a mathematical simulation of the cycle tests, the most suitable pyrite flotation circuit was selected. This flotation circuit was evaluated at an industrial level, obtaining a concentrate with 18.2% copper and a recovery rate of 56.1%. To increase the recovery and the grade of the concentrate, the MAMP plant was modified to include a mill with which to grind the rougher concentrate, and also a flotation column cell as a cleaner stage, obtaining a concentrate with a grade of 21.51% copper and a recovery rate of 88.15%.

Author Contributions: Conceptualization, L.V. and O.P.; methodology, M.S. and J.T.; software, M.S.; validation, L.V., O.P. and M.G.; formal analysis, L.V. and V.R.; investigation, L.V.; resources, M.G. and V.R.; data curation, M.S. and J.T.; writing—original draft preparation, L.V.; writing—review and editing, O.P.; visualization, O.P.; supervision, M.S.; project administration, L.V. and M.G.; funding acquisition, M.G and V.R. All authors have read and agreed to the published version of the manuscript.

Funding: This research received no external funding.

Data Availability Statement: The data are available upon request to the authors. The slag under study was obtained from a specific mining company; therefore, there may be restrictions on providing data other than that published.

Acknowledgments: The authors would like to thank the Department of Metallurgical Engineering of the Faculty of Engineering of the University of Atacama, the National Mining Company, and the

Carola Coemin Mining Group for their support in carrying out this research, and Evelyn Cárdenas for her assistance in translating this article.

Conflicts of Interest: Victor Rivera San Martin from Empresa Nacional de Mineria (ENAMI) and Miguel Gonzalez Troncoso from Planta Cerrillos, part of the Carola Coemin Mining Group, state that the article represents the views of the scientists and not the company.

References

1. Phiri, T.C.; Singh, P.; Nikoloski, A.N. The potential for copper slag waste as a resource for a circular economy: A review–Part II *Miner. Eng.* **2021**, *172*, 107150. [CrossRef]
2. Xu, L.; Liu, Y.; Chen, M.; Wang, N. Efficient Recycling of Valuable Metals from Waste Copper Slag by Using Secondary Aluminum Dross as a Novel Reductant. *Metall. Mater. Trans. B* **2022**, *53*, 2824–2837. [CrossRef]
3. Shen, H.; Forssberg, E. An overview of recovery of metals from slags. *Waste Manag.* **2003**, *23*, 933–949. [CrossRef] [PubMed]
4. Nadirov, R.K. Recovery of valuable metals from copper smelter slag by sulfation roasting. *Trans. Indian Inst. Met.* **2019**, *72*, 603–607. [CrossRef]
5. Mihajlovic, A.; Kamberovic, Z.; Korac, M.; Gavrilovski, M.; Jovanovic, N. The effect of cooling rate of slag from primary copper production in the valorization of copper in the flotation process. *Metall. Mater. Eng.* **2015**, *21*, 127–141. [CrossRef]
6. Voisin, L.; Pizarro, C.; Ossandón, J. Determination of the behaviour of valuable metals during the controlled cooling of copper smelting slags to clarify their recovery by grinding and flotation. In Proceedings of the 5th International Slag Valorization Symposium, Leuven, Belgium, 3–5 April 2017; pp. 205–208.
7. Xue, P.; Li, G.; Qin, Q. Recovery of copper from slow cooled ausmelt furnace slag by floatation. In *Characterization of Minerals, Metals, and Materials*; Springer: Cham, Switzerland, 2016; pp. 621–628.
8. Alexandrova, R.; Paunova, R.; Marinov, M.; Petrov, P. Study on the behavior of flotation fayalite waste from copper production at nonisothermal heating. *J. Chem. Technol. Metall.* **2014**, *49*, 90–93.
9. Das, B.; Mishra, B.K.; Angadi, S.; Pradhan, S.K.; Prakash, S.; Mohanty, J. Characterization and recovery of copper values from discarded slag. *Waste Manag. Res.* **2010**, *28*, 561–567. [CrossRef] [PubMed]
10. Kundu, T.; Senapati, S.; Das, S.K.; Angadi, S.I.; Rath, S.S. A comprehensive review on the recovery of copper values from copper slag. *Powder Technol.* **2023**, *426*, 118693. [CrossRef]
11. Shamsi, M.; Noparast, M.; Shafaie, S.Z.; Gharabaghi, M.; Aslani, S. Effect of grinding time on flotation recovery of copper smelting slags in Bardaskan district. *J. Min. Environ.* **2015**, *6*, 237–249.
12. Roy, S.; Datta, A.; Rehani, S. Flotation of copper sulphide from copper smelter slag using multiple collectors and their mixtures *Int. J. Miner. Process* **2015**, *143*, 43–49. [CrossRef]
13. Shamsi, M.; Noaparast, M.; Seied Ziadin Shafaie, S.Z.; Gharabaghi, M. Synergism effect of collectors on copper recovery in flotation of copper smelting slags. *Geosystem Eng.* **2016**, *19*, 57–68. [CrossRef]
14. Sipunga, E. Optimization of the Flotation of Copper Smelter Slags from Namibia Custom Smelters' Slag Mill Plant. Doctoral Dissertation, University of the Witwatersrand, Johannesburg, South Africa, 2015.
15. Sibanda, V.; Sipunga, E.; Danha, G.; Mamvura, T.A. Enhancing the flotation recovery of copper minerals in smelter slags from Namibia prior to disposal. *Heliyon* **2020**, *6*, e03135. [CrossRef] [PubMed]
16. Linsong, W.; Zhiyong, G.; Honghu, T.; Li, W.; Haisheng, H.; Wei, S.; Yongbao, Q.; Yue, Y. Copper recovery from copper slags through flotation enhanced by sodium carbonate synergistic mechanical activation. *J. Environ. Chem. Eng.* **2022**, *10*, 107671. [CrossRef]
17. Pavez, O.; Herrera, H.; González, M.; Rivera, O. Copper recovery through smelter slag flotation in Atacama, Chile: An industrial case study. *Ing. Investig.* **2021**, *41*, e84162. [CrossRef]
18. Štirbanović, Z.; Urošević, D.; Đorđević, M.; Sokolović, J.; Aksić, N.; Živadinović, N.; Milutinović, S. Application of Thionocarbamates in Copper Slag Flotation. *Metals* **2022**, *12*, 832. [CrossRef]
19. King, J.; Li, Q.; Wang, A.; He, C.; Zhou, J.; Deng, H.; Xu, R. Evaluation of grinding media wear-rate by a combined grinding method. *Miner. Eng.* **2015**, *73*, 39–43. [CrossRef]
20. Valderrama, L.; González, M.; Santander, M.; Zazzali, B. Recuperación de cobre contenido en escoria de cobre mediante flotación *Holos* **2018**, *5*, 40–50. [CrossRef]
21. Guarda, N. Efecto del Enfriamiento Controlado en el Consumo de Energía y la Recuperación de Valiosos en el Procesamiento de Escorias de Fusión de Cobre. Master's Thesis, Universidad de Chile, Santiago, Chile, 2015. Available online: http://repositorio.uchile.cl/handle/2250/137113 (accessed on 25 November 2024).
22. Petrakis, E.; Komnitsas, K. Effect of Grinding Media Size on Ferronickel Slag Ball Milling Efficiency and Energy Requirements Using Kinetics and Attainable Region Approaches. *Minerals* **2022**, *12*, 184. [CrossRef]

23. Rahman, R.M.; Ata, S.; Jameson, G.J. The effect of flotation variables on the recovery of different particle size fractions in the froth and the pulp. *Int. J. Miner. Process.* **2012**, *106–109*, 70–77. [CrossRef]
24. Agar, G.E.; Stratton-Crawley, R.; Bruce, T.J. *Optimizing the Design of Flotation Circuits;* CIM Bulletin: Philadelphia, PA, USA, 1980; pp. 173–180.
25. Tong, X.; Han, B.; Ren, S.P.; Yang, B. Recovery of copper from copper smelter slag by flotation. *Appl. Mech. Mater.* **2014**, *496–500*, 406–409. [CrossRef]

minerals

Review

Ultrasonic Enhancement for Mineral Flotation: Technology, Device, and Engineering Applications

Xiaoou Zhang [1], Huaigang Cheng [1,2,*], Kai Xu [1], Danjing Ding [1], Xin Wang [1], Bo Wang [1] and Zhuohui Ma [1]

1 Institute of Resources and Environmental Engineering, Shanxi University, Taiyuan 030006, China; xiaoouzhang2425@163.com (X.Z.); 20200290401047@email.sxu.edu.cn (K.X.); 13485306952@163.com (D.D.); 15735175936@163.com (X.W.); wangbo@sxu.edu.cn (B.W.); mazhuohui@sxu.edu.cn (Z.M.)

2 College of Chemical Engineering, Qinghai University, Xining 810016, China

* Correspondence: chenghg@sxu.edu.cn; Tel.: +86-351-7018553

Abstract: In the past five years, the number of articles related to ultrasonic mineral flotation has increased by about 50 per year, and the overall trend is on the rise. The most recent developments in ultrasonics for flotation process intensification are reviewed herein, including effects of ultrasound treatment on an aqueous slurry, improvement in flotation methods and technological processes, device development tracking, and application effects in mineral process engineering. At this point in time, there are pilot-scale flotation tests to evaluate the feasibility of ultrasonic pretreatment technology for industrial use to enhance residue flotation separation, and the results showed that the recovery rate of concentrate is increased by about 10%. Four aspects of ultrasonic flotation process improvement are summarized, namely, changing the ultrasonic parameters, the synergistic effect of ultrasound and reagents, the ultrasonic effect of particles with different-sized fractions, and application to new systems. In addition, the effect of ultrasonic flotation mechanisms is explored through a quadratic model and numerical simulation. The combination of ultrasonic flotation with other fields, such as magnetic fields, to enhance the separation efficiency and recovery of minerals is also a future trend. It is also proposed that ultrasonic flotation technology will be used with big data, industrial Internet of Things, and automatic control technology to achieve deep bundling, optimizing the flotation process by implementing remote monitoring and control of the flotation process.

Keywords: ultrasound; flotation; technology; device; enhancement

1. Introduction

In the mineral processing field, flotation has been widely applied as an efficient separation method for the purification and recovery of minerals such as coal gasification fine slag, apatite, dolomite from chalcopyrite, galena, copper oxide, quartz, and carbide slag [1]. However, with the rising prominence of resource depletion and environmental issues, challenges such as low efficiency, environmental pollution, and high energy consumption in conventional flotation have necessitated the development of ultrasonic flotation. Ultrasonic flotation is an innovative physical treatment method that introduces ultrasonic waves into the flotation system, generating a large number of microbubbles and localized high-temperature, high-pressure environments, thereby altering the physicochemical properties of the mineral surfaces. Leveraging the cavitation and acoustic radiation force effects produced by ultrasonic waves promotes the adsorption of particles in the pulp with flotation agents, enhancing flotation efficiency and recovery rates and reducing the use of chemicals and energy consumption.

In the last 20 years, with the continuous development of ultrasonic treatment, ultrasonic flotation devices have also been upgraded. New devices, such as ultrasonic generators and ultrasonic flotation cells, have been gradually applied in industrial production, achieving high efficiency, energy savings, and environmental protection. With continuous exploration and optimization in engineering practice, ultrasonic flotation shows broad application prospects in the treatment of low-grade beneficiation, improving resource utilization and reducing production costs.

Ultrasonic flotation has been reviewed [2–5], and the application of ultrasonic in flotation, leaching, and gravity has also been systematically discussed in the literature [2], with the interaction between ultrasound and minerals and agents focused on in the flotation part. The application of ultrasound in flotation was reviewed in Ref. [3], including sludge removal, oxide film removal, desulfurization, microbubble formation, flotation agent dispersion, and aggregation. The application principle of ultrasonic technology in mineral flotation is discussed in Ref. [4], including the effect of ultrasound on minerals, flotation reagents, and slurries, as well as their application status. The mechanism of ultrasonic treatment in mineral flotation is discussed in detail in Ref. [5], including the transient cavitation effect, stable cavitation effect, and acoustic radiative force effect, as well as the influence of the main parameters of ultrasonic treatment on mineral flotation and the application of ultrasonic treatment in minerals (e.g., cleaning effect, ultrasonic corrosion, and desulfurization), flotation agents (e.g., dispersion and emulsification, and changes in the properties and microstructure of the agent solution), and slurry (e.g., microbubble formation and aggregation).

Effects of ultrasound treatment on an aqueous slurry (the fundamentals of ultrasonic flotation), improvement of flotation methods and processes, development of devices, and application effects in mineral processing engineering are systematically discussed in this paper. According to the different shapes of transducers, the ultrasound devices discussed in Ref. [3] are divided into three types: horn type, plate type, and focusing type. The ultrasonic flotation equipment in Ref. [4] is intended for trough, probe, and ultrasonic flotation columns. Different from the aforementioned literature, this review mainly focuses on the development trajectory of ultrasonic flotation equipment, focusing on the development of ultrasonic flotation equipment in the last five years. At present, the technology of ultrasonic mineral flotation is still in its early stages. By analyzing the development of ultrasonic flotation and exploring its potential applications in different mineral processing methods, this study provides a theoretical basis and practical guidance for further optimization and promotion of the technology. At the same time, future research directions and development trends are proposed to address the problems and challenges of the existing technology.

2. Methodology

2.1. Innovation of This Review

This paper is divided into four parts: effects of ultrasound treatment on an aqueous slurry, improvement of flotation methods and technological processes, device development tracking, and application effects in mineral process engineering. The most notable innovations are primarily concentrated in the second and third parts. The innovation of the second part lies in providing readers or enterprises with ways to obtain better flotation effects by adjusting the ultrasonic flotation parameters, adding some reagents to achieve the synergistic effect of ultrasonic and reagents, changing the particle size of flotation minerals, and adding ultrasound to the flotation system to obtain improved flotation effects. The innovation of the third part lies in systematically summarizing and discussing the development process of mineral ultrasonic flotation device.

2.2. Literature Research

For this review, Web of Science, Elsevier and x-mol were used to screen publications, primarily based on the topics of "ultrasonic flotation", "ultrasonic flotation & mineral", and "ultrasonic flotation device". The references for ultrasonic flotation devices are mostly from patents, and few existing journal articles are dedicated to ultrasonic flotation devices. Figure 1 shows the published literature on these topics in Web of Science. It can be seen from the figure that the current ultrasonic flotation of minerals is still in the early stage, and the quantity of related literature is relatively small. The publications were manually filtered by checking the titles and reading the abstracts. The text on ultrasonic flotation devices is primarily based on patents from the last 20 years, while the remaining content on ultrasonic flotation mainly discusses journal articles from the last five years. Ultimately, 84 articles matching the topics were selected for thorough analysis and summarization.

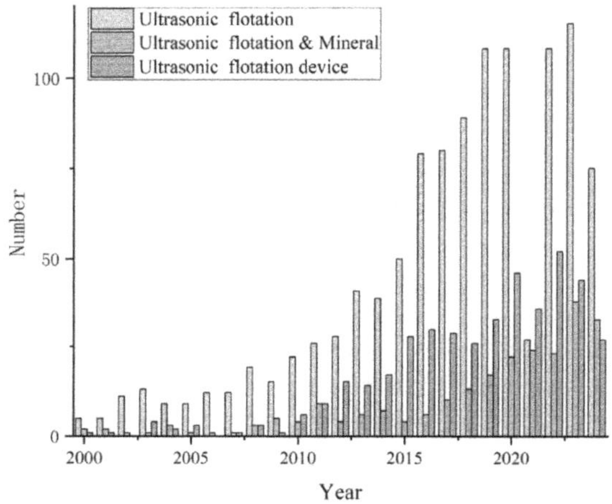

Figure 1. Distribution of published articles since 2000 (to 23 September 2024) from Web of Science.

3. Effects of Ultrasound Treatment on an Aqueous Slurry

The possible effects of ultrasound flotation mainly include the transient cavitation effect, stable cavitation effect, and acoustic radiative force effect. Cavitation and acoustic radiative force are two common effects present in water during ultrasonic treatment, as shown in Figure 2. Transient cavitation requires high sound intensity (greater than 10 W/cm^2) to trigger and undergo a rapid growth and violent collapse process, resulting in the formation of so-called hot spots: local high temperatures and pressures. It has a strong impact and peeling effect on the impurities on the mineral surface and can dissolve some components on the mineral surface. Stable cavitation occurs at low sound intensity, and the bubbles produce stable small-amplitude pulsations, improving the collision efficiency with fine particles, helping to remove soluble impurities from the mineral surface, and dispersing and emulsifying the flotation agent [5]. Siderite is soluble, and ts dissolution process will produce $CO_3{}^{2-}$. $CaCO_3$ is generated by $CO_3{}^{2-}$ and activator Ca^{2+} to occupy the active sites on the surface of quartz, attaching to the quartz surface such that the quartz shows the properties of calcite and reduces its hydrophobicity. Mechanical effects induced by ultrasonic cavitation can clean the surface of minerals and increase their active sites to a certain extent, thus improving their floatability [6]. By way of illustration, regarding the outcome of ultrasonic enhancement of coal gasification fine slag, ultrasonic treatment can enhance the hydrophobicity of the surface of coal gasification fine slag and increase the number of active sites on the surface, thus treating coal gasification fine slag flotation tailings more effectively [7]. In addition, microbubbles generated by ultrasonic cavitation can act as a bridge to promote the adsorption of collecting agent on the mineral surface and enhance mineral flocculation by increasing the contact angle and decreasing the zeta potential [8]. Acoustic radiation force is the movement of particles, oil, or bubbles in the ultrasonic field. Particle surface properties are altered by acoustic radiation forces in ultrasonic flotation to promote coal particle aggregation and flotation [9]. Ultrasonic flotation, under the synergistic effect of cavitation and acoustic radiation force, realizes the improvement of flotation efficiency compared with traditional flotation.

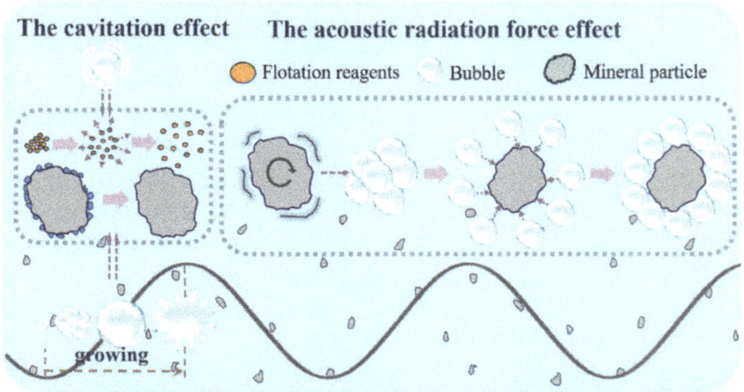

Figure 2. Schematic diagram of cavitation effect and acoustic radiation effect.

4. Improvement of Flotation Method and Technological Process

4.1. Ultrasonic Parameters

Selecting the appropriate ultrasonic parameters for a specified flotation system is the main consideration in the ultrasonic flotation process. The setting of ultrasonic parameters and their types will have a greater or lesser impact on cavitation and acoustic radiation forces, thus affecting the action of particles and bubbles in ultrasonic flotation, as well as the outcome of ultrasonic flotation.

Significant progress has also been made in recent years regarding the search for the optimal ultrasonic frequency for different minerals and flotation conditions. Previous studies have mostly used low-frequency ultrasound, in the range of approximately 20–60 kHz, and most of the ultrasound frequencies used in the references included in this review are in this range [10]. Specifically, innovation in the literature [11] lies in the study of the effects of ultrasonic standing waves at three different frequencies, 80 kHz, 100 kHz and 120 kHz, on the behavior of flotation bubbles and flotation effects, and 100 kHz was identified as the ideal frequency for ultrasonic flotation. Under the action of the 100 kHz ultrasonic standing waves, the bubble aggregates and coal–bubble aggregates were formed within 450 ms and 20 ms, respectively, which significantly improved the flotation efficiency of fine coal. This process is mainly attributed to the high-frequency pressure change generated by the ultrasonic standing wave, which promotes the rapid aggregation of bubbles and coal particles and thus enhances the flotation effect. In addition, the authors also mention that a study shows that 200 kHz ultrasonic standing wave can significantly improve the flotation rate of fine coal particles without the addition of chemicals or collectors. A key innovation of Chen et al.'s study is that they studied bubble–coal particle interactions in a 600 kHz ultrasonic standing wave field (USW), a significant increase compared to the frequency used in conventional ultrasonic flotation techniques (usually in the range of 20–50 kHz). Under the action of 600 kHz USW, stable tiny cavitation bubbles can be created and maintained on the surface of coal particles. The bubbles, with the action of the initial

Bjerknes force, migrate to the low-pressure area (the minimum point of pressure), causing coal particles to converge at these low-pressure points. In addition, the secondary Bjerknes force establishes an attraction between small bubbles and more atmospheric bubbles, which enhances the adhesion between coal particles and bubbles. In contrast to conventional interactions, particles can be attracted and collected by the bubbles without having to undergo collision and attachment processes. This efficient bubble–particle interaction mechanism significantly improves flotation recovery [12]. The innovation of the above two papers lies in the use of high-frequency ultrasonic standing waves to enhance the collision and aggregation efficiency of mineral particles.

Regarding the effect of ultrasonic power on ultrasonic flotation, Kang et al. studied the effect of ultrasonic pretreatment on graphite flotation effect and explored the effects of three different ultrasonic powers of 0.8 kw, 1.2 kw, and 1.6 kw on the concentrate mineral yield, and the results showed that the higher the ultrasonic power, the higher the concentrate yield [13]. Esmeli et al. found that in the lignite flotation process, the power of ultrasound as a pretreatment is in the range of 30–60 W, and the recovery rate of combustible components increases with the increase in power [14]. Deng et al. studied the effects of ultrasonic pretreatment at 120 W, 180 W, 240 W, and 300 W on the quartz flotation effect and found that with an increase in ultrasonic power, the recovery rate of quartz sand concentrate was the highest at 240 W and the impurity removal effect was good, but the quartz flotation was not further improved at 300 W [15]. High power and prolonged use of ultrasound may adversely affect the conditioning process by creating turbulent conditions and/or increasing the temperature in the conditioning medium [16]. For potassium salts in materials with different particle sizes, with the increase in ultrasonic power in the range of 15–70 W, the recovery rate of coarse materials decreases and the recovery of intermediate materials first increases and then decreases, while the selectivity of flotation can be improved for fine-grade materials [17].

The ultrasound type contains a standing wave field in addition to the traditional diffuse field. In an ultrasonic field, both particles and bubbles are subjected to the force of acoustic radiation. The acoustic radiation force effect is related to the movement of bubbles or particles. In addition to Jin et al. and Chen et al. [11,12], Chen et al. [3] also used ultrasonic standing waves to study the effects of acoustic radiation forces at 50 kHz, where particles and oil droplets can be precisely manipulated [18]. Conversely, the acoustic radiation force on the bubble is sensitive to low-frequency ultrasound, and it decreases dramatically when the bubble size differs significantly from the resonance radius.

To sum up, the development of ultrasonic flotation frequencies in recent years has primarily focused on selecting different ultrasonic frequencies for the flotation system to determine the optimal ultrasonic flotation frequency, with the frequencies tested in the above cases exceeding the typical ultrasonic frequency range of 20–60 kHz. Future development trends may involve experimenting with high-frequency ultrasonic flotation systems to improve the efficiency of ultrasonic selection. Regarding the development of ultrasonic flotation power, it has been found that the flotation effect is not always positively correlated with the ultrasonic power, and the same power may also produce different effects on the same ore sample of different particle sizes. Regarding the development of ultrasonic types, in recent years, ultrasonic flotation has begun to gradually adopt the standing wave field, which is no longer limited to the traditional diffusion field, and the application range of the ultrasonic standing wave field in ultrasonic flotation may gradually expand in the future.

4.2. The Synergistic Effect of Ultrasound and Reagents

Ultrasound can produce a synergistic effect with flotation reagents (such as collecting agents, frothing agents, etc.) to improve the flotation efficiency. Ultrasound can enhance the adsorption of the collecting agent on the surface of minerals, as well as improving the stability and selectivity of bubbles. The research on the synergistic effect between ultrasound and reagents has also made significant progress in recent years.

Firstly, ultrasound can enhance the flotation efficiency by directly affecting the physical properties of the flotation reagents. For instance, what is innovative about Zhang et al.'s study [19] is that they used ultrasound to modify starch, introducing ultrasonic treatment to enhance the inhibition of starch in hematite reverse flotation, where starch acted as an inhibitor to make the surface of iron oxides hydrophilic, and as a flocculating agent to make the fine iron oxides form larger flocs. The ultrasonic treatment carried out prior to flotation enhanced the inhibition of hematite and selective flocculation while improving starch (MS) solubility and increasing the content of straight chain starch. This helped to reduce hematite entrainment losses and increase the iron grade of the concentrate.

Secondly, ultrasound can influence the effect of flotation reagents on minerals. For example, the novelty of [20] lies in the study of the inhibition effect on galena when ultrasonic treatment is used in combination with sulfuric acid, and the results indicate that ultrasound can significantly improve the inhibition ability of sulfuric acid on galena. The ultrasonic-induced cavitation effect creates localized hot spots on the mineral surface. They deliver an intensified energy input, which expedites the chemical interaction with sulfuric acid on the galena surface. This accelerated reaction fosters the precipitation of $PbSO_4$, thereby enhancing the selective separation process between galena and chalcopyrite. In addition, the separation of galena and chalcopyrite using sulfuric acid requires high temperatures (60–80 °C) and long processing times (more than 25 min), whereas ultrasound can be used to reduce the reaction time to 6 min at room temperature, significantly reducing energy consumption and improving separation efficiency.

Thirdly, ultrasound is able to affect the surface adsorption of flotation reagents on minerals, enhance the adsorption capacity of surfactants, and promote the desorption of surfactants from the surface of mineral residues [7]. A particularly innovative aspect of the research by Huang et al. is that they study the effects of ultrasonic pretreatment on the surface dissolution of scheelite as well as the physicochemical properties and microstructure of sodium oxalate solution. They found that ultrasonic pretreatment of sodium oleate or scheelite promoted the adsorption of sodium oleate on the surface of scheelite, which increased the heat of adsorption and thermal reaction rate of sodium oleate and scheelite, as well as the hydrophilicity of the surface, and led to an improvement in the flotation rate and recovery of scheelite. Nevertheless, ultrasonic treatment is more suitable for sodium oleate and less effective for scheelite [21]. Moreover, regarding coal flotation, one of the innovations of Sun et al. lay in the use of ultrasonic desorption to remove surfactant from the surface of gangue, which reduced the hydrophobicity of gangue and enhanced the wettability difference between coal and gangue. The second innovation lay in the discovery that the addition of Ca^{2+} promotes the adsorption of surfactant on the organic matter; however, the effect on the gangue was not significant, which helped to increase the flotation selectivity. Through the synergistic effect of ultrasonic treatment and Ca^{2+}, the flotation response of low-rank coal was improved, and better flotation performance was obtained [22].

To sum up, development in the synergistic effect of ultrasound and reagents in recent years mainly includes the modification of flotation reagents by ultrasound, the combined benefits of flotation agents and ultrasonic processing, and the realization of surface modification of minerals through ultrasonic flotation. Applications include galena, scheelite, and coal. The foci of research include exploring the optimal combination of different types and concentrations of reagents and ultrasound and how they work together to modify mineral surfaces.

4.3. Particle Size Fractions

With regard to the influence of particle size fractions on the effect of ultrasonic flotation, some progress has been made in recent years. For instance, Chu et al. [23] innovatively applied ultrasonic pretreatment technology to improve the flotation performance of spodumene, conducted a systematic study on the effect of ultrasonic pretreatment on spodumene with different particle sizes, and revealed the effect of particle size on the effect of ultrasonic pretreatment, which is less often involved in traditional flotation studies. The quantity of NaOL adsorbed on the mineral surface was noticeably increased by ultrasonic pretreatment; thus, the floatability of spodumene was further enhanced. For coarse particles, ultrasonic pretreatment was more effective compared to conventional methods in optimizing the surface physicochemical properties. Conversely, for fine particles, ultrasonic pretreatment increased the amount of dissolution of the mineral, but did not show a significant advantage over the conventional method. Filippov et al. investigated the role of ultrasonic treatment in improving the flotation selectivity of specific minerals (sylvite KCl and halite NaCl), which had not been extensively explored in previous studies. The combined effect of ultrasonication on the physicochemical parameters of the slurry related to the flotation behavior of particles of different sizes was investigated for a potash ore flotation system, and it was found that the use of ultrasonic treatment can be beneficial for the treatment of fine-grained ore and reduce the loss of potash in the process of desliming [17]. Mao et al. innovated by investigating the effect of ultrasonic pretreatment technology on the improvement of the flotation performance of lignite particles of different size fractions, and found that the coarse particles of lignite showed a significant increase in the recovery of combustible matter after ultrasonic pretreatment, while the recovery of fine particles decreased. In this process, ultrasonic treatment mainly improved the flotation efficiency by crushing the minerals, and particles that were too fine were not conducive to lignite flotation [24]. Chen et al. determined that the sulfur content of a 0.019 mm residue increased after sonication by analyzing the residue before and after sonication [25]. Wang et al. also determined through flotation that ultrasonic flotation has a more significant crushing effect on fine slag particles of entrained-flow gasification coal fine slag than traditional flotation, and that the flotation effect is better [26].

In recent years, in order to improve the effect of mineral flotation, the influence of particle size on the effect of ultrasonic flotation has been increasingly considered. Regarding the impact of particle size fractions on research progress in ultrasonic flotation, the flotation objects have expanded to include spodumene, potash ore, lignite, and gasification coal fine slag flotation systems. For different flotation systems, the effects of coarse and fine particle size effects vary. The impact of ultrasonic flotation on specific particle sizes needs to be specifically analyzed, with universal conclusions remaining lacking. This analysis may be conducted by considering the mineral composition of each particle size and determining the particle size for flotation accordingly. Alternatively, flotation could be carried out separately for each particle size, with the most suitable flotation particle size being judged according to the flotation effect. Currently, the flotation fractions mainly comprise 60 mesh, 100 mesh, 200 mesh, 400 mesh, and 900 mesh.

4.4. Application System

With a better understanding of the mechanisms of ultrasonic flotation, new flotation systems are being developed to accommodate different types of minerals and more complex ores.

The trend of improvement in ultrasonic flotation lies primarily in its application in existing flotation systems. Kruszelnicki et al.'s innovation was the first systematic exploration of the role of ultrasonic pretreatment in the flotation of carbonaceous copper-bearing shale. The specific mechanism of the effect on the floatability of the minerals was explored in depth, specifically on the generation of ultrafine bubbles and the cleaning effect through transient bubble collapse. Ultrafine bubbles contributed to the enhancement of the adhesion efficiency of minerals to the bubbles, improving the flotation selectivity. The findings showed that the ultrasonic pretreatment improved the flotation performance of minerals by altering the charge state of the mineral surfaces and enhancing the adsorption of the collecting agent. Under the ultrasonic pretreatment condition, the amount of collecting agent to achieve the same flotation recovery was significantly reduced [27]. Liao et al. were innovative in applying ultrasonic pretreatment technology to the separation process of pyrrhotite and chlorite. It was observed that ultrasonic pretreatment could promote the dissolution of iron oxide particles on the surface of pyrrhotite and the desorption of iron ions adsorbed on the surface of chlorite in the mixed minerals, which in turn facilitated the separation between pyrrhotite and chlorite [28].

Ultrasonic treatment also involves the selective separation of minerals through the application of ultrasonic flotation. For instance, in an innovative move, Wu et al. applied ultrasonic treatment to selectively recover ilmenite from titanaugite [29]. The floatability of titanaugite was suppressed, while the floatability of ilmenite was improved. Selective separation was enhanced when the ultrasonic treatment time was 5 min and the power was 500 W. The innovation of Fang et al. was that they investigated how ultrasonic pretreatment affects the selective adsorption of ilmenite and its associated minerals into sodium oleate through surface dissolution. A new way to physically improve mineral surface properties to enhance the flotation performance of specific minerals was provided. Ultrasound promotes the oxidation of Fe^{2+} to Fe^{3+} and the formation of $Fe(OH)_3$ precipitates that readily adsorbs NaOL, increasing ilmenite recovery while reducing the adsorption of accompanying minerals for effective separation [30].

New advances made in recent years regarding the application system of ultrasonic flotation include the expansion of the system for the flotation of carbon-bearing copper shale, pyrrhotite, ilmenite, and so on. Ultrasonic treatment can significantly improve the flotation effect of carbon-bearing copper shale, promote the separation of pyrite and chlorite, and increase the flotation of ilmenite by inhibiting the flotation of titanaugite, realizing the selective separation of the two. It can also play a positive role in the selective flotation of ilmenite from gangue minerals of ilmenite.

Regarding improvements in the ultrasonic flotation process, the main development trends are outlined in Figure 3, in terms of ultrasonic parameters, ultrasonic frequency, ultrasonic power, type of ultrasound affect cavitation, and acoustic radiation force effects, which in turn affect the ultrasonic flotation effect. The main trends include selecting the appropriate ultrasonic frequency within higher frequencies beyond the low-frequency range, selecting the appropriate ultrasonic power, and trying to use the ultrasonic standing wave for ultrasonic flotation. These approaches aim to enhance flotation efficiency through the interaction of ultrasound and flotation reagents. Additionally, the impact on different particle sizes of mineral particles varies. With a deeper understanding of the mechanisms of ultrasonic flotation, new flotation systems are being developed to adapt to different types of minerals and more complex ores.

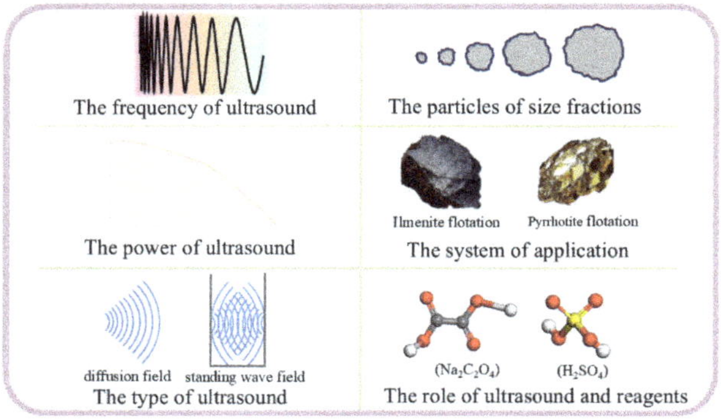

Figure 3. Development trends in the ultrasonic flotation process.

5. Ultrasonic Flotation Device

The design and optimization of ultrasonic flotation devices are key to achieving high-efficiency flotation. In this chapter, the development track of ultrasonic flotation devices is discussed, and the development trends of ultrasonic flotation devices are predicted.

Ordinary ultrasonic devices and ultrasonic flotation devices have some fundamental similarities in composition, but there are clear differences in their components and design features due to their different application purposes. Ordinary ultrasonic devices mainly consist of ultrasonic generator, ultrasonic transducer/probe, power amplifier, control system, and cooling system. The ultrasonic generator is responsible for generating electrical signals to control the ultrasonic waves, while the ultrasonic transducer is responsible for converting these electrical signals into mechanical vibration. The ultrasonic transducer/probe will then convert electrical energy into ultrasonic vibrations. Compared with ordinary ultrasound, the ultrasonic flotation device features more flotation columns and an ultrasonic vibration plate acting on the pulp and bubbles.

The development of ultrasonic flotation devices in the early stages primarily focused on the ultrasonic device itself, such as enhancing the ultrasonic device and its mobility [31], achieving high-precision control of the ultrasonic vibrator [32], making the ultrasonic generator probe more stable to increase the ultrasonic wave output for floating [33], and improving the rigidity of the ultrasonic flotation device [34]. With the development of ultrasonic flotation devices, research has mainly focused on enhancing the ultrasonic flotation effect in various respects, including pulp flow [35–38], cleaning the surface of mineral particles [39–47], ultrasonic emulsification [40,48–50], and flotation bubbles [27,34,41–45]. Some ultrasonic flotation devices for laboratory use have also been invented, allowing for better pulp dispersion [51] or for the ultrasonic intensity to be adjusted according to the experimental needs [52].

Ultrasonic devices may not act directly in the flotation process. As a case in point, the innovation of Liao et al. is the use of an ultrasonic nebulizer instead of a direct drug feeder, which improves the phosphorite flotation efficiency and concentrate indices by increasing the contact area and improving the mass transfer efficiency. The medicament disperser and phosphorus slurry disperser utilize an annular tube structure with an additional packing layer to promote more thorough mixing and contact, further improving the phosphorus ore flotation efficiency. The main device includes a raw ore pulp tank, flotation cells, a pharmaceutical dissolution tank, and an ultrasonic atomization drug feeder [53].

In recent years, there has been development in ultrasonic flotation devices that can effectively clean mineral surfaces. An innovative approach by Chen et al. introduced ultrasonic pretreatment to effectively clean the graphite surface and dissociate the graphite intergrown with impurities. The main device includes an ultrasonic device, flotation cells and agitators, which utilize the cavitation effect to produce tiny bubbles that bridge fine graphite particles to achieve effective separation [35]. The research demonstrates that the intensity of ultrasonic cavitation has a significant effect on graphite recovery: both horn-type and bath-type ultrasonic waves contribute to flotation. Horn-type ultrasonic waves show a more pronounced effect, leading to an increase of 7% in the recovery of the flotation, whereas bath-type ultrasonic waves resulted in only a 2% increase [10]. Ma et al. were innovative in inventing an ultrasonic cleaning structure that disperses and strips mineral particles through ultrasonic action to effectively reduce slurry viscosity and inter-coagulation, improve the hydrophobicity and capture rate, reduce mineral encapsulation, and enhance bubble transport, thus improving the sorting efficiency and concentrate grade. This device is suitable for granular minerals with a diameter range of 100–400 μm [54]. The innovation of Long et al. lies in the coal slurry flotation pretreatment device integrating a classification device and an ultrasonic generator, which utilizes the cavitation effect of ultrasonic waves to make the fine sludge on the surface of the fine-grained coal slurry after classification come off and effectively improves the quality of the finedcoal after flotation. The main device includes a pretreatment device, dosing device, classification device. and flotation device. Another advantage of the device is its ability to process both fine- and coarse-grained slime over a wide range of sizes [43].

Regarding ultrasonic emulsification of flotation reagents, Yin et al. invented a device for emulsifying flotation reagents, which resulted in particle sizes of approximately 0.2–10 μm through ultrasonic emulsification [48]. Li et al. were innovative in developing a device for improving the flotation separation efficiency of mixed materials from the positive and negative electrodes of waste lithium batteries. Their device utilizes ultrasound to sufficiently emulsify the reagent, forming a fine homogeneous shape that improves the adsorption capacity on the surface of graphite negative electrode materials, enhancing the flotation separation efficiency. The main device includes flotation cells, ultrasonic tanks, ultrasonic components, impellers, and transfer pumps [54]. The innovative aspect of Yin et al.'s work is the achievement of synergistic coupling between ultrasonic emulsification and ultrasonic intensification, whereby ultrasonic emulsification enables the collecting agent and foaming agent to form an emulsion with the water and improves their dispersion in the water and ultrasonic intensification effectively destroys the structure of the fly ash and facilitates the exposure and separation of the carbon. The primary device comprises screening machinery, an ultrasonic generator, and flotation cells. By improving the efficiency of char removal and reducing the risk of environmental pollution, the efficient reduction and resource utilization of fly ash was achieved. The patent clarifies that the ultrasonic frequency during emulsification treatment is 32–48 kHz, the ultrasonic frequency is set to 28–48 kHz during enhanced flotation, air is introduced in the bubbling flotation process, the control ventilation is 0.05–0.10 m^3/h, and the emulsion reagent is added to the mortar five times, each addition following a ratio of 1.4 L of mortar to 100 mL of emulsifying reagent [55].

The ultrasonic flotation device has seen significant progress and innovation in enhancing the ultrasonic flotation effect. Gao et al.'s innovative approach involves the use of focusing action-type ultrasonic reinforcement to enhance ultrasonic effects, facilitate the interaction between bubbles and minerals, and augment the efficiency of collisional adhesion. The flotation column is narrowed from the middle to the top and bottom, and a total of six ultrasonic transducers are set on the side walls of the flotation column to strengthen the collision and adhesion efficiency of flotation bubbles and mineral particles [56]. Sun's team successively researched the ultrasonic frequency range and action area and developed a multi-frequency ultrasonic flotation device [57] and inclined ultrasonic flotation device [58].

The innovation of the former lies in the use of multi-frequency ultrasonic excitation mechanism, which applies multi-level ultrasonic action on the flotation bubbles to strengthen the collision–adhesion effect between bubbles and mineral particles. The innovation of the latter is that the ultrasonic transducer is arranged in a V shape with the horizontal plane to enlarge the area of ultrasonic transducer action such that the flotation bubbles in the slurry are more fully subjected to ultrasonic action.

New achievements have also been made in the development of ultrasonic bubble refining technology for ultrasonic flotation devices. Shi et al.'s invention of an ultrasonic jet coupling device refines bubbles with ultrasonic waves. This enhances the adsorption of bubbles to mineral particles in the pulp, forming a swirling pulp circulation field, which accelerates the discharge of gangue minerals and improves the efficiency of coal slurry flotation [59]. On the other hand, Wei et al. invented an ultrasonic vibration device that allows for the control of both the speed and volume of bubble generation [60]. The device facilitates the aggregation of bubbles under the action of ultrasonic waves, with the bubble size increasing with an increase in the thickness of the vibration plate of the ultrasonic transducer [61].

Regarding the combined effect of the ultrasonic flotation device and sieve plate, the innovation lies in the combination of the ultrasonic transducer and filling-type sieve plate. This combination facilitates the removal of the mineral surface cover layer and the dispersion of mineral particles through ultrasonics, while simultaneously achieving static stratification via the filled screen plate. As a result, the hydrophobic particles adhering to the bubbles are less likely to be dislodged, strengthening the sorting effect and preventing the screen plate from being clogged. The main device includes a flotation column, ultrasonic power supply, foam overflow vessel, concentrate outlet, filling type sieve plate and ultrasonic transducer [62]. Li et al. prevented the settling of coarse particles by filling the sieve plate and effectively eliminated the effect of ultrasonic damping pulsation on the stability of the bed layer, achieving effective recovery of coarse particles of 0.25–1 mm [50].

In addition, the recent development of big data, the industrial Internet of Things and automatic control technology has made significant progress and applications in the field of flotation. Rockwell Automation provides comprehensive industrial data analysis solutions to help mining companies optimize the flotation process [63]. Siemens' MindSphere is an open Internet of Things operating system that enables remote management and optimization [64]. ABB provides automatic control systems; for example, it can be used in mineral processing equipment to improve productivity through precise control [65]. FL Smidth developed advanced control algorithms for optimizing mineral processing processes, including flotation operations [66]. Li et al. invented an ultrasonic flotation device that incorporates a PID intelligent control system. This addition enables the device to achieve intelligent fluidization, intelligent tailing, and intelligent ultrasound, effectively improving the sorting accuracy and adaptability of the device, thereby improving the flotation efficiency [50]. With the industrial application of ultrasonic flotation technology, the combined utilization of ultrasonic flotation equipment and other technologies can significantly enhance flotation efficiency. The utilization of big data, Internet of Things, and automatic control technologies represents key trends for future development.

Regarding the development of ultrasonic flotation devices, there has been a progression, as shown in Figure 4, from the early simple design of ultrasonic flotation equipment, such as a movable part consisting of a floating slide to expand the area of the floating guiding surface [31], to the ultrasonic flotation process, such as an increase in the mixing degree of the slurry, to the ultrasonic flotation of minerals, such as cleaning the surface of the mineral particles [39], and then to the ultrasonic flotation of flotation equipment to achieve an emulsifying effect on the flotation reagents [48]. At present, flotation equipment is increasingly relying on the advancements in big data, the Internet of Things, automation, and other technologies. The development trend of ultrasonic flotation equipment, along with the deep integration of these three technologies, will likely become a key trend in future development. In addition, the combination of ultrasonic fields and other physical fields such as magnetic fields [67] and electric fields in ultrasonic flotation equipment is also a future development trend aimed at achieving efficient flotation of complex minerals.

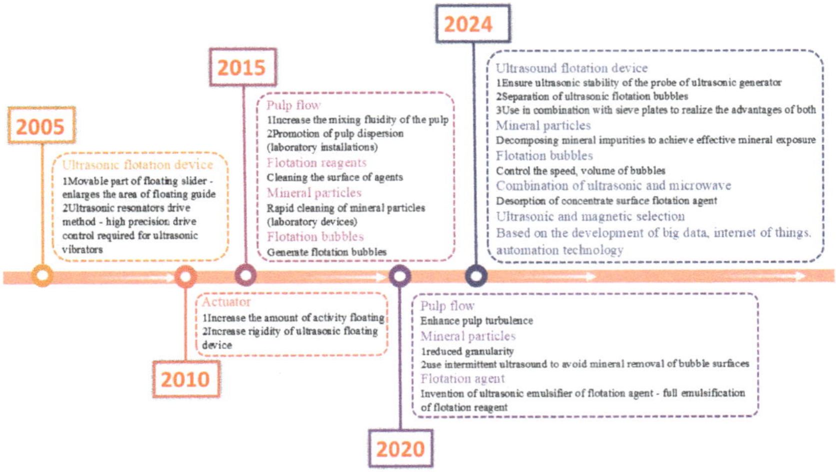

Figure 4. The development track of ultrasonic flotation devices.

6. Application Effects in Mineral Process Engineering

This chapter summarizes the literature in terms of effects and products. There are several ways to improve the flotation effect, mainly involving altering roughness of minerals, cleaning surfaces of minerals, increasing porosity of minerals, promoting dispersion of minerals, increasing the active site of minerals, and changing the zeta potential of minerals, as shown in Figure 5. In general, ultrasonic flotation enhances mineral flotation by acting on the mineral surfaces, but the existing literature does not demonstrate that ultrasonic flotation has an effect on the crystal structure of minerals.

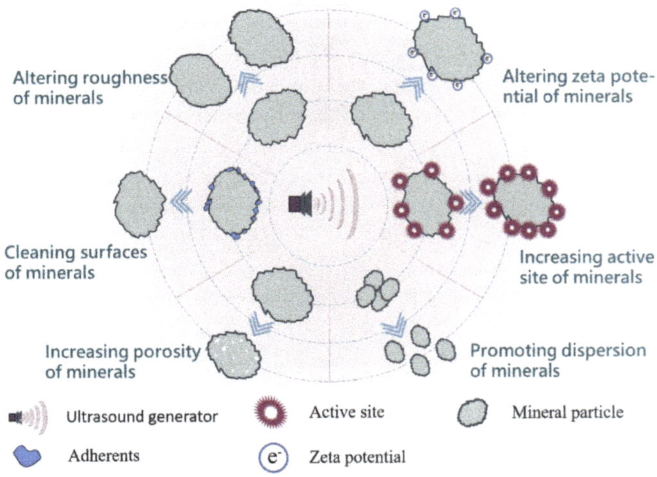

Figure 5. Ways to improve the flotation effect.

6.1. Pretreatment

Ultrasonic flotation pretreatment can improve the flotation efficiency by promoting mineral dispersion and cleaning the mineral surface through ultrasound. For example, Gong et al. removed clay from the surface of foundry dust coal powder with the help of an ultrasound-assisted flotation process, which is innovative in the recycling and utilization of foundry dust as it targets a specific type of waste, in this case foundry dust. It was found that after ultrasonic pretreatment for 30 min, the yield was 55.3%, the loss on ignition (LOI) was 19.3%, and the recycled aggregate concrete was 69.22% [68]. In a follow-up study, the recovery rate of pulverized coal in tailings was increased to 74.36% after pre-soaking-assisted mechanical stirring pretreatment was used to replace ultrasonic flotation, which shows that although ultrasonic pretreatment can improve the flotation effect, it may not be the best pretreatment method [69]. The innovation of the study by Wang et al. lies in the application of ultrasonic flotation to the recovery of graphite and cathode materials from spent lithium-ion batteries (LIBs), a relatively new field of application, where conventional flotation is seldom involved in the recovery of such materials. It was also found that ultrasonic treatment could effectively destroy the organic binder and electrolytes on the surface of electrode materials in waste LIBs, thus enhancing the flotation differentiation between graphite and cathode materials. Additionally, graphite and cathode materials in LIBs were effectively separated under a dosage of methyl isobutyl methanol (MIBC) of 400 g/ton and diesel fuel of 200 g/ton and 3-min ultrasonic treatment [70]. Liao et al. were innovative in applying ultrasonic pretreatment to the separation process of pyrite and chlorite by promoting the dissolution of iron oxide particles on the surface of pyrite and the desorption of adsorbed iron ions on the surface of chlorite through ultrasonic treatment, which in turn enhanced the separation of pyrite and chlorite by increasing the recovery of pyrite from 77.50% to 81.72% and decreasing the content of magnesium oxide from 6.57% to 5.74%, thus increasing the grade and recovery of pyrite [16]. Chen's team were innovative in introducing ultrasound to the graphite recovery process of spent carbon cathodes from the electrolytic aluminum industry, using ultrasound to promote particle dispersion to clean the mineral surface and increase the porosity of the graphite surface, thus enhancing the flotation performance. The slurry was pretreated with ultrasound for 30 min. The slurry recovery was 96.24%, which was 10.2% higher than that of the conventional flotation, the tailings content was 9.68%, which was 17.31% lower than that of the conventional flotation, and the grade of the concentrate was 84.95% [71]. The innovation of Bagheri et al. lies in the use of an ultrasonic pretreatment technique to improve the flotation performance of sphalerite in their study, whereby ultrasonic treatment removed the impurities and oxide layer from the mineral surface and cleaned the mineral surface, improving the hydrophilicity of the sphalerite and enhancing the interaction between the collecting agent and mineral surface. The sphalerite recovery was enhanced from 70.30% to 73.79%. However, the ultrasonic treatment also increased the flotation of unwanted associated minerals (e.g., pyrite), with the flotation enhancement of tailings pyrite increasing from 31.95% to 54.38%, resulting in a significant decrease in the separation efficiency of zinc [72]. This indicates that ultrasonic pretreatment does not always have a positive effect on mineral flotation.

It is also worth noting that Dongfang et al. were innovative in investigating the effect of ultrasonic pretreatment on the flotation behavior of galena, not only for a single mineral sample but also for a mixture of galena and pyrite. It was found that the recovery of galena increased after the ultrasonic pretreatment of a single galena sample; however, the recovery of galena decreased after the ultrasonic pretreatment of a mixture of galena and pyrite samples [73]. This suggests that an interfering effect may occur between minerals when ultrasonic flotation of multiple minerals is performed.

Ultrasonic flotation pretreatment can also improve flotation efficiency by acting on the mineral surface and changing the mineral surface structure or properties. For example, Mao et al. were innovative in studying how ultrasonic pretreatment changed the surface topography of lignite, including the generation of cracks, gaps, and cavities, and how these changes affected the flotation performance of lignite. It was found that an increase in ultrasonic power and treatment time led to an increase in lignite surface roughness and a decrease in concentrate yield by step, except at 18 watts. In particular, for lignite flotation treated with 108 W ultrasonic waves for 3 min, the concentrate yield was reduced by 16.65% compared to untreated lignite flotation, while the measurements of surface roughness and induction time provided a new technical tool for evaluating the effect of ultrasonic pretreatment [74]. Xue et al. were innovative in applying ultrasonic pretreatment to the hydrophobic–hydrophilic separation (HHS) process in order to enhance residual carbon in coal gasification fine slag (CGFS) enrichment and revealed the adsorption behavior of n-heptane on the residual carbon surface by molecular dynamics simulations, which is typically a spontaneous process driven by van der Waals forces. This simulation approach provides micro-level insights for understanding the ultrasonic pretreatment affecting HHS processes. It was found that ultrasonic pretreatment plays a positive role in the effectiveness of HHS by changing the surface properties of CGFS particles and enhancing the hydrophobicity of residual carbon through mechanical action and chemical oxidation. In addition, the microbubbles generated by ultrasound helped the particles to trap and separate at the oil–water interface, which further improved the efficiency of HHS. With the continuous increase in ultrasound power, the ash content of the concentrate was reduced from 31.81% to 18.93%, while the ash content of the tailings was greater than 96.00% [75]. Chen et al. dissociated elemental sulfur and silver-bearing lead minerals in the oxidative pressure leaching residue of zinc sulfide through the use of ultrasonic waves, increasing the surface structural differences between the elemental sulfur and the silver-bearing lead minerals, which contributed to their separation. The innovation lies in the evaluation of the feasibility of ultrasonic pretreatment technology for enhancing the flotation separation of residues in industry through pilot-scale flotation tests. It was found that the sulfur grade and concentrate recovery were increased by 11.63% and 9.10%, respectively, and that the recovery of lead and silver in the tailings was increased by 8.72% and 9.50%, respectively [25]. Ann Bazar et al. found that ultrasonic pretreatment improved the dispersion of talc by altering the surface properties of talc and enhanced the adsorption of inhibitors by generating active sites on the surface of talc. This was shown to reduce the flotation recoveries of talc by approximately 15% at lower inhibitor dosages and 8% at higher inhibitor dosages [76]. Ren et al. combined ultrasonic pretreatment slurry and composite collector flotation to enrich residual carbon from coal gasification slag. The structural properties and surface chemistry of CGFS were improved, and the ultrasonic pretreatment slurry demonstrated an excellent crushing effect on tightly connected carbon-gray particles. The loss indices of three concentrates obtained by combined processes compared to traditional flotation increased by 1.52 times, 1.42 times, and 1.20 times, respectively, while the loss indices of tailings decreased by 51.48%, 86.46%, and 35.58%, respectively [77].

The above cases show that in recent years, ultrasonic flotation pretreatment has been demonstrated to improve the flotation efficiency by promoting the dispersion of minerals, cleaning the surface of minerals, acting on the surface of minerals, and changing the structure or properties of the surface of minerals. However, during the ultrasonic flotation of multiple minerals, interactions between minerals may form, resulting in poor flotation results. Ultrasonic flotation pretreatment does not always improve mineral flotation.

6.2. Simultaneous Treatment

Cleaning the mineral surface is the main method of simultaneous ultrasonic flotation to improve flotation efficiency. For example, the innovative nature of [78] lies in the application of ultrasonic treatment during different stages of flotation. The ultrasonic treatment exposed more adsorption sites by having a cleansing or stripping effect on the mineral surfaces, thereby increasing the specific surface area, and also altered the zeta potential of the mineral surfaces, making them more negatively charged, which contributed to the enhancement of the sodium oleate adsorption capacity on the mineral surface. Simultaneous ultrasonic treatment improved the flotation separation efficiency of magnesite and dolomite, resulting in a flotation recovery of 91.06% for magnesite concentrate, nearly a 20% increase compared to conventional flotation. Chen et al. innovatively enhanced the efficient recovery of valuable metals from waste printed circuit boards (WPCBs) by introducing ultrasound into the flotation process and found that ultrasonic flotation facilitated the detachment of epoxy resin and minute metal fragments from the glass fiber surfaces in WPCBs. Through the cavitation effect and surface cleaning, the hydrophobicity of the non-metallic components and particle dispersion were enhanced, which facilitated a significant increase in the total metal recovery, as well as Cu, Al, and Zn recovery from the WPCBs. They were increased by 7.20%, 4.82%, 15.41%, and 5.58%, respectively [79].

Moreover, simultaneous ultrasonic flotation treatment utilizing cavitation and acoustic radiation forces goes beyond cleaning mineral surfaces. For example, one of the innovations of Mao et al.'s study on the effect of ultrasonic flotation in the pulp–froth zone on the selectivity and kinetics of flotation of high-ash lignite was the elucidation of the mechanism of ultrasonic flotation in the froth zone. The ultrasonic treatment carried out in the froth zone produced large bubbles and a thin froth layer due to ultrasonic cavitation and oscillation, improved the water-carrying recovery of fine particles, increased the concentrate yield by 14.53% and the combustible recovery by 21.42%, and concentrate ash was reduced by only 1.56%. The ultrasonic flotation in the pulp zone provided optimal surface cleaning of the coal surface, where the ultrasonic treatment increased the concentrate yield by 9.72% and the combustible recovery by 18.54%, while the ore ash was reduced by 7.75% [80].

On the other hand, acoustic radiation force effect plays a greater role in high-frequency simultaneous ultrasonic treatment. For example, Chen et al. innovated by studying the effects of different ultrasonic standing wave (USW) frequencies on the formation of microbubbles, dispersion of conventional flotation bubbles (CFBs), and movement of particles in the fine coal flotation process. The research demonstrated that the high-frequency USW produced more microbubbles as carriers of coal particle motion and promoted the collision and adhesion between coal particles and bubbles, increasing the recovery of fine-grained materials and flotation rate through acoustic radiative force. The maximum recovery obtained by the 600 kHz USW increased from 22% in the absence of USW to 53%, while low-frequency (50 kHz) USW may reduce these metallurgical indicators [81].

Nevertheless, simultaneous ultrasonic flotation treatment does not always enhance the flotation efficiency. Take the innovative application of heterogeneous carrier flotation technology as an example, using Cu^{2+}-treated coarse-grained pyrite as a carrier to enhance fine-grained chalcopyrite recovery. Although ultrasonic treatment may disrupt the Cu^{2+}-treated layer on the surface of pyrite, ultrasonic treatment does not significantly promote the loss of chalcopyrite fines due to the hydrophobic forces between chalcopyrite and pyrite, whereas acidic treatment effectively separates chalcopyrite fines on coarse pyrite particles. At pH 2, approximately 96% of chalcopyrite fines were separated from Cu^{2+}-treated coarse pyrite particles [82].

Ultrasonic pretreatment and ultrasonic simultaneous treatment may have a more positive effect on flotation efficiency. For example, Wang et al.'s study introduced a novel approach that comprehensively applied ultrasonic pretreatment and ultrasonic emulsification technology. They investigated the effects of ultrasonic pretreatment slurry and ultrasonic emulsification flotation agents on flotation enrichment of residual carbon in entrained-flow gasification coal fine slag. The effect of ultrasonic flotation on the crushing of entrained-flow gasification coal slag particles was more significant than that of traditional flotation. Ultrasonic treatment enhanced the dissociation of gasified coal fines through mechanical and cavitation effects, removed the high-ash fines on the surface, and reduced the mechanical entrainment of high-ash fines, thus improving the selectivity of flotation. In addition, the ultrasonic emulsification improved the dispersion of the collecting agent on the surface of the coal dust gasification fines and reduced the surface tension, which helped the collecting agent cover the particle surface more uniformly and enhanced the hydrophobicity in the flotation process, thus improving the flotation efficiency of the fines. The concentrate yield and ash content were reduced by 9.94% and 16.54%, respectively, while the quality of concentrate improved significantly, with the flotation perfection index increasing by 12.60% [26].

Ultrasonic technology not only improves the hydrophobicity of minerals but also proves its versatility in solid waste resource utilization by reducing the ash content of concentrates and increasing the recovery of tailings ash and combustibles. Ultrasonic power and treatment time are key parameters that affect the effectiveness of ultrasonic treatment. The above shows that an increase in ultrasonic power can significantly reduce the ash content of the concentrate, but care also needs to be taken to avoid over-treatment leading to reoxidation of mineral particles or other adverse effects. Moreover, the ultrasonic standing wave (USW) is used as a fine-grained flotation aid, and its frequency has a significant effect on flotation efficiency. The overall flotation effects of ultrasonic flotation are shown in Table 1.

In addition, ultrasonic treatment technology as a physical treatment method has obvious advantages in reducing the use of chemical additives and lowering the environmental impact, which helps to promote the development of the mineral processing industry in a green and sustainable direction. Even so, it is not a complete replacement for traditional flotation; rather, it serves as a complementary means to improve the overall flotation performance by improving mineral surface properties and promoting particle dispersion. The joint application of ultrasonic technology with other technologies, such as synergizing with composite capture agents, provides new ideas for improving the flotation efficiency of specific minerals. This integrated and innovative approach may become a significant direction for the future development of ultrasonic flotation technology.

With the current findings, ultrasound technology shows wide applicability in different types of mineral and material recovery and performs well in improving flotation efficiency, but the differences in the effect of ultrasound treatment on flotation performance at different frequencies and treatment times indicate that there is scope for further optimization of ultrasound parameters to achieve optimum flotation results for different types of minerals. However, as mineral properties, particle size, chemical composition, and other factors increase, they may negatively affect the recovery rate and selectivity of the final destination minerals and require targeted optimization. The introduction of ultrasonic treatment technology may increase energy consumption and operating costs; therefore, evaluating its economic efficiency and cost/benefit ratio is crucial for realizing industrial applications.

Table 1. Mineral ultrasonic flotation effects.

Process-Specific		System	Comparison of Results Compared to Conventional Flotation
Ultrasonic pretreatment	Coal	Residual carbon from CGFS	Keeping the ash fluctuations of tailings under 2%, the concentrate ash saw a drop from 31.81% to 18.93% as ultrasonic power escalated.
			The combined method increased the three concentrates' LOIs by 1.52, 1.42, and 1.20 times, and LOIs of cut tailings by 51%, 86%, and 36%.
		Combustible components	The presence of micro–nano bubbles during 37 kHz and 80 kHz treatments raised the combustible recovery by roughly 35% and 25%, respectively.
	Sphalerite	Low ash, fine lignite	As ultrasonic power and time rose, concentrate yield generally fell, with the exception at 18 W.
	High-zinc-grade tailing	Silver-bearing and sulfur lead minerals	Sulfur grade rose 11.63% and concentrate recovery increased by 9.10%, while the lead and silver recoveries from tailings climbed to 8.72% and 9.50%, respectively.
	Chlorite	Sphalerite	Flotation recovered over 73% of Zn from tailings, yielding a rougher concentrate at 28.61% Zn grade.
	Spent carbon cathode	Pyrrhotite	Pyrrhotite recovery rose from 77.50% to 81.72%, while the MgO content dropped from 6.57% to 5.74%.
	Foundry dust	Graphite	Flotation recovery rates reached 96.24%, tailings contained 9.68%, and concentrate grades peaked at 84.95%.
	Gasification coal fine slag	Clay minerals and coal powder	Tailings yield was 55.3%, with 19.3% LOI and 69.22% RPC.
Simultaneous ultrasonic treatment	Lignite	Residual carbon	Concentrate yield dropped by 9.94%, ash content dropped by 16.54%, and the flotation perfect index rose by 12.60%.
	Magnesite and dolomite	Concentrates and combustibles	In pulp-zone ultrasonic flotation, concentrate yield and combustible recovery rose 9.72% and 18.54%, respectively, while froth zone saw gains of 14.53% and 21.42%.
	Waste printed circuit boards	Magnesite and dolomite	Magnesite concentrate recovery hit 91%.
	Sphalerite	Pyrite particles	Metal recovery from WPCBs collectively rose by 7.20%, with specific gains of 4.82% in Cu, 15.41% in Zn, and 5.58% in Al.

7. Research Prospect and Direction

Physical and chemical reactions in the ultrasonic flotation process can be better understood and predicted using quadratic models and numerical simulations. This helps to optimize ultrasonic flotation process parameters, improve beneficiation efficiency, and reduce the energy consumption and environmental impact. Experimental data from ultrasonic flotation technology can also be used to validate and improve the quadratic model, aligning it more with practical applications. Jiang et al. developed a quadratic model to predict polyvinyl chloride purity, explored the interactions between pH, $AlCl_3$ concentration, and ultrasonic treatment time, and achieved microplastic recovery and purity of more than 99.65% [83]. Xue et al. verified the spontaneous adsorption process of n-heptane on the residual carbon surface in coal gasification fine slag through molecular dynamic simulation, which was driven by van der Waals forces and provided a theoretical basis for improving hydrophobic–hydrophilic separation efficiency through ultrasonic pretreatment [75].

The use of composite flotation technology combining ultrasound with other physical fields (e.g., electric fields and magnetic fields) and the development of new ultrasound-assisted chemical treatment methods also represent a development trend. Wang et al. combined ultrasonic slurry strong magnetic flotation technology to realize the separation and extraction of carbon and iron powder in the gangue while enhancing the efficiency of decarbonization and iron removal [49]. Cao et al. found that the combination of ultrasound and electrochemistry for the desulfurization of fine coal was more effective than the ultrasonic enhanced flotation method [84]. Future development trends may involve more research and practical applications that combine ultrasonic flotation with magnetic fields, electric fields, and other force fields to enhance beneficiation efficiency, reduce ore waste, and lower production costs. Additionally, it is important to explore the synergistic mechanisms of ultrasonic flotation with these force fields, optimize process parameters, and develop new equipment and devices. This will help promote the development of ore extraction and separation technology and realize more effective utilization of mineral resources.

8. Conclusions

Ultrasonic flotation technology, a new type of mineral processing technology, has played a significant role in improving the utilization rate of mineral resources and reducing environmental pollution, such as coal, tailing containing zinc mineral, pyrite, potash ore, spent lithium-ion batteries, and electronic waste.

In recent years, the development of the flotation method has focused on four main areas: changing the ultrasonic parameters, the synergistic effect of ultrasonic waves and reagents, the ultrasonic effect of different particle sizes, and the application of new systems. These foci aim to enhance the efficiency of flotation through the use of ultrasonic flotation technology. Additionally, ultrasonic flotation devices have evolved from simple designs to the integration of complex technologies. The evolution of equipment for ultrasonic flotation has progressed from early designs aimed at improving efficiency by increasing the area of the floating guide surface to more advanced technologies that enhance slurry mixing, clean the surfaces of mineral particles, and emulsify flotation reagents. Current developments rely on big data, the Internet of Things, automation, and other technologies, reflecting future trends. Additionally, the integration of ultrasonic fields with other physical fields aims to increase the flotation efficiency of complex minerals. Regarding the effectiveness of ultrasonic flotation, its applicability across different types of minerals and material recovery is broad; however, performance varies greatly with different frequencies and treatment times, influenced by factors such as mineral characteristics, particle size and chemical composition. Furthermore, ultrasonic flotation is being combined with other physical fields (magnetic, electric, etc.) to improve the flotation of complex minerals. Research employing quadratic models and numerical simulations is underway to study ultrasonic flotation mechanisms and further investigate these mechanisms.

Future research should focus on gaining an in-depth understanding of the cavitation and acoustic radiation effects generated by ultrasonic flotation. This can be achieved through the application of ultrasonic treatment in the flotation process to optimize operating parameters and integrate other technologies. The goals include facilitating the crushing or aggregation of minerals, cleaning or modifying mineral surfaces, enhancing the synergistic effects of composite reagents, aiming to improve mineral flotation effectiveness, and promoting the resourceful use of waste minerals. Ultrasonic flotation facilitates green flotation, high-efficiency flotation, and intelligent flotation.

Author Contributions: Writing—original draft preparation, X.Z.; writing—review and editing, X.Z.; supervision, H.C.; Language polishing, X.W., B.W., Z.M.; project administration, K.X., D.D.; funding acquisition, H.C. All authors have read and agreed to the published version of the manuscript.

Funding: This work was financially supported by the National Key Research and Development Plan of China (2022YFB4102102), Centrally Guided Local Science and Technology Development Fund Projects of Qinghai Province (2024ZY006), National Natural Science Foundation of China (U20A20149, 22478232), Qinghai Province Basic Research Plan (2023-ZJ-920M), and Shanxi University Interdisciplinary Construction Project. We are grateful for their financial support and valuable guidance.

Data Availability Statement: Data are contained within the article.

Conflicts of Interest: The authors declare no conflicts of interest.

References

1. Zhou, Y.; Wang, X.; Liu, H.; Cheng, Y.; Ma, P. Carbide Slag Comprehensive Utilization Device. C.N. Paten CN111717929-A, 29 September 2020. Available online: https://pss-system.cponline.cnipa.gov.cn/conventionalSearch (accessed o 24 September 2024).
2. Huang, W.; Hu, Y.; Huang, L.; Zeng, X.; Zhang, T. Application Progress of Ultrasonic Technology in Mineral Processing. *App Chem. Ind.* **2020**, *49*, 3172–3177.
3. Chen, Y.; Truong, V.N.T.; Bu, X.; Xie, G. A Review of Effects and Applications of Ultrasound in Mineral Flotation. *Ultrason Sonochem.* **2020**, *60*, 104739. [CrossRef] [PubMed]
4. Chen, L.; Lu, D.; Wang, Y.; Cheng, Z. Research and Application of Ultrasonic Technology in Mineral Flotation and Developmen Trend. *Chin. J. Nonferrous Met.* **2021**, *31*, 1042–1056.
5. Zhang, H.; Du, M.; Hu, H.; Zhang, H.; Song, N. A Review of Ultrasonic Treatment in Mineral Flotation: Mechanism and Recen Development. *Molecules* **2024**, *29*, 1984. [CrossRef] [PubMed]
6. Yin, W.; Cai, L.; Ma, Y.; Wang, Y. Mechanism of Ultrasonic Cavitation to Improve the Effect of Siderite on Quartz Flotatior *Physicochem. Probl. Miner. Process.* **2023**, *59*, 165930. [CrossRef]
7. Jiao, Y.; Yang, Z.; Han, X.; Wang, K.; Fang, C.; Zhao, Z.; Tang, W. Ultrasonication Improves the Flotation of Coal Gasification Fin Slag Residue. *Minerals* **2024**, *14*, 363. [CrossRef]
8. Ozkan, A.; Esmeli, K. Use of Ultrasonic Treatment as a Pre-Phase in the Shear Flocculation Process. *Ultrasonics* **2023**, *134*, 10705 [CrossRef]
9. Yang, A.; Wang, L. A Fundamental Study of Flotation Separation of Mineral Particles Using Ultrasound-Induced Bubbles. *Mine Eng.* **2024**, *207*, 108573. [CrossRef]
10. Zhou, S.; Tong, Z.; Dong, L.; Bu, X.; Ni, C.; Xie, G.; Alheshibri, M. A Comparative Study on the Influence of Single and Combine Ultrasounds Assisted Flake Graphite Flotation. *Ultrason. Sonochem.* **2023**, *99*, 106551. [CrossRef]
11. Jin, L.; Wang, W.; Tu, Y.; Zhang, K.; Lv, Z. Effect of Ultrasonic Standing Waves on Flotation Bubbles. *Ultrason. Sonochem.* **2021** *73*, 105459. [CrossRef] [PubMed]
12. Chen, Y.; Ni, C.; Xie, G.; Liu, Q. Toward Efficient Interactions of Bubbles and Coal Particles Induced by Stable Cavitation Bubble under 600 kHz Ultrasonic Standing Waves. *Ultrason. Sonochem.* **2020**, *64*, 105003. [CrossRef]
13. Kang, W.; Li, H. Effect of Ultrasonic Pretreatment on Cleaner Flotation of Graphite. *J. China Univ. Min. Technol.* **2020**, *49*, 1193–119
14. Esmeli, K. The Influence of Ultrasound Treatment on the Hydrophobic Flocculation of Lignite. *Int. J. Coal Prep. Util.* **2024**, *44*, 1–18 [CrossRef]
15. Deng, J.; Luo, D.; Bai, J.; Rong, K. Effect of Ultrasonic Pretreatment on Flotation Purification of Quartz. *Physicochem. Probl. Mine Process.* **2024**, *60*, 190008. [CrossRef]
16. Gungoren, C.; Baktarhan, Y.; Demir, I.; Ozkan, S.G. Enhancement of Galena-Potassium Ethyl Xanthate Flotation System by Lov Power Ultrasound. *Trans. Nonferrous Met. Soc. China* **2020**, *30*, 1102–1110. [CrossRef]
17. Filippov, L.O.; Filippova, I.V.; Barres, O.; Lyubimova, T.P.; Fattalov, O.O. Intensification of the Flotation Separation of Potash Ore Using Ultrasound Treatment. *Miner. Eng.* **2021**, *171*, 107092. [CrossRef]
18. Luo, X.; Cao, J.; Gong, H.; Yan, H.; He, L. Phase Separation Technology Based on Ultrasonic Standing Waves: A Review. *Ultrason Sonochem.* **2018**, *48*, 287–298. [CrossRef]
19. Zhang, M.; Xu, Z.; Wang, L. Ultrasonic Treatment Improves the Performance of Starch as Depressant for Hematite Flotation *Ultrason. Sonochem.* **2022**, *82*, 105877. [CrossRef]
20. Cao, Q.; Li, Y.; Zou, H.; Liu, D.; Yan, Y.; Zhang, H. Synergistic Effects of Ultrasonication on the Flotation Separation of Galena from Chalcopyrite Using Sulfuric Acid as a Depressant. *Min. Eng.* **2024**, *206*, 108501. [CrossRef]
21. Huang, Z.; Kuang, J.; Yuan, W.; Yu, M.; Wang, X. Regulation Mechanism of Ultrasonication on Surface Hydrophobicity of Scheelite. *Colloid Surf. A* **2021**, *629*, 127412. [CrossRef]
22. Sun, X.; Zhang, L.; Xie, Z.; Li, B.; Liu, S. Improvement of Low-Rank Coal Flotation Based on the Enhancement of Wettability Difference between Organic Matter and Gangue. *J. Surfactants Deterg.* **2021**, *24*, 269–279. [CrossRef]

23. Chu, H.; Chen, L.; Lu, D.; Wang, Y.; Zheng, X. Ultrasonic Pretreatment of Spodumene with Different Size Fractions and Its Influence on Flotation. *Ultrason. Sonochem.* **2022**, *82*, 105889. [CrossRef] [PubMed]
24. Mao, Y.; Xia, W.; Peng, Y.; Pan, L.; Xie, G. Impact of Ultrasonic Pretreatment on the Flotation of Lignite Particles of Different Size Fractions. *Int. J. Coal Prep. Util.* **2022**, *42*, 315–330. [CrossRef]
25. Chen, C.; Zhang, S.; Liu, Q.; Chen, L.; Xian, Y.; Wang, Y. Ultrasonic Pretreatment for Enhancing Flotation Separation of Elemental Sulfur and Silver-Bearing Lead Minerals from an Oxidative Pressure Leaching Residue of Zinc Sulfide. *Min. Eng.* **2024**, *205*, 108495. [CrossRef]
26. Wang, W.; Liu, D.; Tu, Y.; Jin, L.; Wang, H. Enrichment of Residual Carbon in Entrained-Flow Gasification Coal Fine Slag by Ultrasonic Flotation. *Fuel* **2020**, *278*, 118195. [CrossRef]
27. Kruszelnicki, M.; Hassanzadeh, A.; Legawiec, K.J.; Polowczyk, I.; Kowalczuk, P.B. Effect of Ultrasound Pre-Treatment on Carbonaceous Copper-Bearing Shale Flotation. *Ultrason. Sonochem.* **2022**, *84*, 105962. [CrossRef]
28. Liao, Y.; Zhao, G.; Feng, B.; Yan, H.; Wu, H.; Hu, W.; Zhu, D.; Qiu, T. Application of Ultrasonic Pre-Treatment for Flotation Separation Pyrrhotite from Chlorite. *Colloid Surf. A* **2023**, *669*, 131507. [CrossRef]
29. Wu, H.; Fang, S.; Shu, K.; Xu, Y.; Wang, Z.; Luo, L.; Yang, J.; Xu, L. Selective Flotation and Adsorption of Ilmenite from Titanaugite by a Novel Method: Ultrasonic Treatment. *Powder Technol.* **2020**, *363*, 38–47. [CrossRef]
30. Fang, S.; Xu, L.; Wu, H.; Xu, Y.; Wang, Z.; Shu, K.; Hu, Y. Influence of Surface Dissolution on Sodium Oleate Adsorption on Ilmenite and Its Gangue Minerals by Ultrasonic Treatment. *Appl. Surf. Sci.* **2020**, *500*, 144038. [CrossRef]
31. Fujinaga, T.; Yamashita, Y. Ultrasonic Flotation Apparatus Includes Movable Portion Comprising Floating Sliders. J.P. Patent JP2005231746-A, 2 September 2005. Available online: https://www.j-platpat.inpit.go.jp (accessed on 24 September 2024).
32. Fujinaga, T.; Yamashita, Y. Ultrasonic Vibrator Drive Method in Ultrasonic Flotation Apparatus. J.P. Patent JP2005238144-A, 8 September 2005. Available online: https://www.j-platpat.inpit.go.jp (accessed on 24 September 2024).
33. Fujinaga, T.; Yamashita, Y. Ultrasonic Floatation Apparatus Used for Actuator. J.P. Patent JP2006230148-A, 31 August 2006. Available online: https://www.j-platpat.inpit.go.jp (accessed on 24 September 2024).
34. Fujinaga, T.; Yamashita, Y. Ultrasonic Floatation Apparatus for Actuator. J.P. Patent JP2006247641-A, 21 September 2006. Available online: https://www.j-platpat.inpit.go.jp (accessed on 24 September 2024).
35. Hu, Y.; Liu, J.; Sun, W. Flotation Column for Performing Flotation Separation on Nickel Molybdenum Rough Concentrate. C.N. Patent CN102266823-A, 7 December 2011. Available online: https://pss-system.cponline.cnipa.gov.cn/conventionalSearch (accessed on 24 September 2024).
36. Zhu, J. Ultrasonic High-Frequency Vibration Gangue Flotation Chute. C.N. Patent CN204602389-U, 2 September 2015. Available online: https://pss-system.cponline.cnipa.gov.cn/conventionalSearch (accessed on 24 September 2024).
37. Li, C.; Wang, Y.; Ding, Z.; Wang, L.; Yan, X.; Wang, A.; Cao, Y. Cyclone-Static Micro-Bubble Flotation Column Ultrasonic-Strengthened Pipe Flow Section Mineralization Device. C.N. Patent CN205868558-U, 11 January 2017. Available online: https://pss-system.cponline.cnipa.gov.cn/conventionalSearch (accessed on 24 September 2024).
38. Bastug, T. Device for Performing Flotation and Particle Size Reduction in Ores of e.g., Gold. T.R. Patent TR201618296-U, 23 January 2017. Available online: https://www.turkpatent.gov.tr/en/research (accessed on 24 September 2024).
39. Wang, K.; Ceng, D.; Wang, L. Ultrasonic Experiment Single Slot Flotation Machine for Use in e.g., Mine. C.N. Patent CN202741249-U, 20 February 2013. Available online: https://pss-system.cponline.cnipa.gov.cn/conventionalSearch (accessed on 24 September 2024).
40. Qian, L. Flotation Tailing Material for Infrared Dryer Used in Mine Selecting Operation. C.N. Patent CN204594163-U, 26 August 2015. Available online: https://pss-system.cponline.cnipa.gov.cn/conventionalSearch (accessed on 24 September 2024).
41. Yu, L.; Liu, Q.; Gao, Y. Intermittent Fine Particle Oxide Ore Ultrasonic Floatation Device. C.N. Patent CN206613613-U, 7 November 2017. Available online: https://pss-system.cponline.cnipa.gov.cn/conventionalSearch (accessed on 24 September 2024).
42. Sun, X.; Chen, S.; Xuan, X.; Yang, Z.; Wang, Y.; Wang, M. Cavitation and Foaming Integrated Tailing Flotation Device. C.N. Patent CN112495592-A, 16 March 2021. Available online: https://pss-system.cponline.cnipa.gov.cn/conventionalSearch (accessed on 24 September 2024).
43. Long, D.; Wang, B.; Wang, M.; Chen, Z.; Han, G.; Miao, X.; Dong, H.; Yuan, Z.; Gao, Y.; Li, Q. Coal Slime Flotation System Device. C.N. Patent CN213792209-U, 27 July 2021. Available online: https://pss-system.cponline.cnipa.gov.cn/conventionalSearch (accessed on 24 September 2024).
44. He, W.; Lv, Q. Ultrasonic Oscillation Type Nano-Crystal Extracting and Separating Device. C.N. Patent CN114522443-A, 24 May 2022. Available online: https://pss-system.cponline.cnipa.gov.cn/conventionalSearch (accessed on 24 September 2024).
45. Hu, Y.; Sun, W.; Peng, T.; Wang, C.; Wu, S.; Wang, J.; Tao, L.; Gao, Z. Desorption Method of the Concentrate Surface Flotation Reagent Used in Mineral Flotation Technology Field for Mineral Sorting. C.N. Patent CN115041304-A, 13 September 2022. Available online: https://pss-system.cponline.cnipa.gov.cn/conventionalSearch (accessed on 24 September 2024).

46. Liao, X.; Chen, J. Removing Arsenic by Ultrasonic Assisted Flotation of Stannite Tailings. C.N. Patent CN117065938-A, 17 November 2023. Available online: https://pss-system.cponline.cnipa.gov.cn/conventionalSearch (accessed on 24 September 2024).
47. Zhang, Y.; Ma, W.; Chen, X.; Chen, Z. Recycling Graphite from Waste Cathode by Ultrasonic Auxiliary Flotation. C.N. Patent CN117259021-A, 22 December 2023. Available online: https://pss-system.cponline.cnipa.gov.cn/conventionalSearch (accessed on 24 September 2024).
48. Yin, D. Flotation Agent Ultrasonic Emulsifier. C.N. Patent CN208494042-U, 15 February 2019. Available online: https://pss-system.cponline.cnipa.gov.cn/conventionalSearch (accessed on 24 September 2024).
49. Kang, G.; Jiang, M.; Kang, Y.; Deng, Y. Ultrasonic Flotation Pretreatment Device. C.N. Patent CN215465205-U, 11 January 2022. Available online: https://pss-system.cponline.cnipa.gov.cn/conventionalSearch (accessed on 24 September 2024).
50. Li, Y.; Wang, L.; Xia, Y.; Zhang, Y.; Ding, S.; Cao, Y.; Gui, X.; Xing, Y.; Xu, M. Ultrasonic and Mechanical Damping Block Coupling Coarse Particle Flotation Device. C.N. Patent CN114713380-A, 8 July 2022. Available online: https://pss-system.cponline.cnipa.gov.cn/conventionalSearch (accessed on 24 September 2024).
51. Cui, R.; Li, M.; Wang, X.; Xu, H.; Huang, Y. Hanging Slot Type Flotation Machine for Use in Laboratory. C.N. Patent CN204746622-U, 11 November 2015. Available online: https://pss-system.cponline.cnipa.gov.cn/conventionalSearch (accessed on 24 September 2024).
52. Feng, Y.; Lu, K.; Han, B.; Song, B.; Chen, Z.; Ma, T.; Zhao, T.; Cui, L.; Dai, S.; Wang, E. Laboratory Ultrasonic Flotation Machine for Use in Ore Dressing Device Technology Field. C.N. Patent CN114602660-A, 10 June 2022. Available online: https://pss-system.cponline.cnipa.gov.cn/conventionalSearch (accessed on 24 September 2024).
53. Liao, Q.; Xu, F.; Wang, X.; Xiao, L.; Jin, Y.; Tian, C.; She, L.; Xu, S.; Zhang, X.; Fang, C.; et al. Device for Increasing Flotation Efficiency of Phosphorite by Using Ultrasonic Wave Atomizing Administration. C.N. Patent CN209901504-U, 7 January 2020. Available online: https://pss-system.cponline.cnipa.gov.cn/conventionalSearch (accessed on 24 September 2024).
54. Ma, M.; Li, S.; Yan, T.; Fan, Z.; Deng, L. Ultrasonic Cleaning Based Flotation Device. C.N. Patent CN210815740-U, 23 June 2020. Available online: https://pss-system.cponline.cnipa.gov.cn/conventionalSearch (accessed on 24 September 2024).
55. Yan, J.; Li, X.; Huang, Q.; Yu, H.; Chen, J.; Fu, C.; Lin, X. Performing Fly Ash Flotation Coupled with Ultrasonic Emulsification and Ultrasonic Enhanced Adsorption Decarbonization. C.N. Patent CN116441059-A, 18 July 2023. Available online: https://pss-system.cponline.cnipa.gov.cn/conventionalSearch (accessed on 24 September 2024).
56. Gao, K.; Liu, H.; Sun, L.; Zeng, Q.; Chen, S.; Wang, X.; Meng, F.; Wei, H.; Zhang, Z. Focusing Action Type Ultrasonic Reinforced Flotation Bubble Mineralization Device. C.N. Patent CN113102117-A, 13 July 2021. Available online: https://pss-system.cponline.cnipa.gov.cn/conventionalSearch (accessed on 24 September 2024).
57. Sun, L.; Liu, H.; Gao, K.; Zeng, Q.; Chen, S.; Wei, H.; Zhang, Z.; Wang, X.; Meng, F. Multi-Frequency Ultrasonic Action Flotation Bubble Strengthening and Mineralizing Device. C.N. Patent CN113102120-A, 13 July 2021. Available online: https://pss-system.cponline.cnipa.gov.cn/conventionalSearch (accessed on 24 September 2024).
58. Sun, L.; Liu, H.; Gao, K.; Zeng, Q.; Chen, S.; Zhang, Z.; Wang, X.; Meng, F.; Wei, H. Inclined Ultrasonic Flotation Bubble Strengthening and Mineralizing Device. C.N. Patent CN113102119-A, 13 July 2021. Available online: https://pss-system.cponline.cnipa.gov.cn/conventionalSearch (accessed on 24 September 2024).
59. Lin, Z.; Zhang, Y.; Yin, J.; Zhu, H.; Zhu, J.; Shi, Q. Intermittent Ultrasonic Jet Coupling Reinforced Coal Slurry Flotation Device. C.N. Patent CN116786277-A, 22 September 2023. Available online: https://pss-system.cponline.cnipa.gov.cn/conventionalSearch (accessed on 24 September 2024).
60. Wei, H.; Wang, X.; Zong, W.; Sun, L.; Zhang, Z.; Gao, K. Ultrasonic Action Type Bubble Generating Device for Flotation Mineralization Device. C.N. Patent CN116618185-A, 22 August 2023. Available online: https://pss-system.cponline.cnipa.gov.cn/conventionalSearch (accessed on 24 September 2024).
61. Gao, K.; Liu, H.; Sun, L.; Zhang, Z. Effect of Gas Input Conditions and Ultrasound on the Dynamic Behavior of Flotation Bubbles. *Acs Omega* **2022**, *7*, 22326–22340. [CrossRef] [PubMed]
62. Qiu, Y.; Mao, S.; Zhang, T.; Li, X. Filling Type Flotation Column of Ultrasonic Vibration Sieve Plate for Separating Micro-Particle Mineral. C.N. Patent CN219631566-U, 5 September 2023. Available online: https://pss-system.cponline.cnipa.gov.cn/conventionalSearch (accessed on 24 September 2024).
63. Smart Manufacturing Industrial Automation | Rockwell Automation | US. Available online: https://www.rockwellautomation.com/en-us.html (accessed on 24 September 2024).
64. Siemens MindSphere Overview ZH Wp_tcm60-29087.Pdf. Available online: https://www.plm.automation.siemens.com/media/global/zh/Siemens%20MindSphere%20Overview%20ZH%20wp_tcm60-29087.png (accessed on 12 August 2024).
65. MineOptimize. Available online: https://new.abb.com/mining/mineoptimize (accessed on 24 September 2024).
66. Complete Digital Optimisation & Control. Available online: https://www.flsmidth.com/en-gb/company/sustainability/the-missionzero-mine/complete-digital-optimisation-control (accessed on 24 September 2024).

67. Shang, Y.; Bai, J.; Duan, P.; Bai, D.; Duan, D.; Wang, C. Ultrasonic Ore Slurry Strong Magnetic Flotation Carbon Iron Device of Coal Gangue. C.N. Patent CN220759547-U, 12 April 2024. Available online: https://pss-system.cponline.cnipa.gov.cn/conventionalSearch (accessed on 24 September 2024).
68. Gong, X.; Jiang, W.; Hu, S.; Yang, Z.; Liu, X.; Fan, Z. Comprehensive Utilization of Foundry Dust: Coal Powder and Clay Minerals Separation by Ultrasonic-Assisted Flotation. *J. Hazard. Mater.* **2021**, *402*, 124124. [CrossRef]
69. Gong, X.; Hu, S.; Liu, X.; Yang, M.; Jiang, W.; Fan, Z. Flotation Separation of Coal Dust from Foundry Dust Enhanced by Pre-Soaking Assisted Mechanical Stirring. *J. Environ. Manag.* **2023**, *339*, 117899. [CrossRef] [PubMed]
70. Wang, Y.; Tu, Y.; Xu, Z.; Zhang, X.; Chen, Y.; Yang, E. A Promising Method for Recovery of Graphite and Cathode Materials from Spent Lithium-Ion Batteries. *Ionics* **2022**, *28*, 2603–2611. [CrossRef]
71. Chen, X.; Chen, Z.; Zhang, Y.; Chen, X.; Wang, R.; Wu, D.; Ma, W. Study on the Recovery of Graphite from Spent Carbon Cathode via Ultrasound-Assisted Foam Flotation. *J. Clean. Prod.* **2024**, *443*, 141149. [CrossRef]
72. Bagheri, B.; Vazifeh Mehrabani, J.; Farrokhpay, S. Recovery of Sphalerite from a High Zinc Grade Tailing. *J. Hazard. Mater.* **2020**, *381*, 120946. [CrossRef] [PubMed]
73. Dongfang, L.; Chen, L.; Ma, Y.; Zheng, X.; Wang, Y. Effects of Ultrasonic Pretreatment on the Flotation Behavior of Galena with and without the Presence of Pyrite. *Physicochem. Probl. Miner. Process.* **2020**, *56*, 611–624. [CrossRef]
74. Mao, Y.; Xie, G.; Qi, X.; Peng, Y. Effects of Ultrasonic Pretreatment on Particle Size and Surface Topography of Lignite and Its Relationship to Flotation Response. *Energ. Source Part A* **2021**, *43*, 1274–1282. [CrossRef]
75. Xue, Z.; Gao, F.; Dong, L.; Bao, W.; Wang, J.; Fan, P. Promotion of Hydrophobic-Hydrophilic Separation of Coal Gasification Fine Slag through Ultrasonic Pre-Treatment. *J. Environ. Chem. Eng.* **2023**, *11*, 110653. [CrossRef]
76. Ann Bazar, J.; Rahimi, M.; Fathinia, S.; Jafari, M.; Chipakwe, V.; Chehreh Chelgani, S. Talc Flotation—An Overview. *Minerals* **2021**, *11*, 652. [CrossRef]
77. Ren, L.; Gong, Y.; Guo, Q.; Yu, G. Enhanced Flotation of Coal Gasification Fine Slag Based on Ultrasonic Pretreatment Pulp and Composite Collector. *Powder Technol.* **2023**, *424*, 118530. [CrossRef]
78. Yin, W.; Wang, Y.; Ma, Y.; Chen, K. Effects of Ultrasonic Treatment on the Flotation Behavior of Magnesite and Dolomite in a Sodium Oleate System. *Green Smart Min. Eng.* **2024**, *1*, 76–84. [CrossRef]
79. Chen, L.; He, J.; Zhu, L.; Yao, Q.; Sun, Y.; Guo, C.; Chen, H.; Yang, B. Efficient Recovery of Valuable Metals from Waste Printed Circuit Boards via Ultrasound-Enhanced Flotation. *Process Saf. Environ.* **2023**, *169*, 869–878. [CrossRef]
80. Mao, Y.; Bu, X.; Peng, Y.; Tian, F.; Xie, G. Effects of Simultaneous Ultrasonic Treatment on the Separation Selectivity and Flotation Kinetics of High-Ash Lignite. *Fuel* **2020**, *259*, 116270. [CrossRef]
81. Chen, Y.; Chelgani, S.C.; Bu, X.; Xie, G. Effect of the Ultrasonic Standing Wave Frequency on the Attractive Mineralization for Fine Coal Particle Flotation. *Ultrason. Sonochem.* **2021**, *77*, 105682. [CrossRef]
82. Bilal, M.; Park, I.; Ito, M.; Hassan, F.U.; Aikawa, K.; Jeon, S.; Hiroyoshi, N. Carrier Flotation Using Coarse Pyrite for Improving the Recovery of Finely Ground Chalcopyrite: Development of Post-Process of Carrier Flotation to Separate Finely Ground Chalcopyrite Particles from Coarse Pyrite Particles. *Minerals* **2023**, *13*, 916. [CrossRef]
83. Jiang, H.; Zhang, Y.; Wang, C.; Wang, H. A Clean and Efficient Flotation towards Recovery of Hazardous Polyvinyl Chloride and Polycarbonate Microplastics through Selective Aluminum Coating: Process, Mechanism, and Optimization. *J. Environ. Manag.* **2021**, *299*, 113626. [CrossRef]
84. Cao, D.; Xu, X.; Jiang, S. Ultrasound-Electrochemistry Enhanced Flotation and Desulphurization for Fine Coal. *Sep. Purif. Technol.* **2021**, *258*, 117968. [CrossRef]

minerals

MDPI

Review

Advances and Prospects on Flotation Enhancement of Difficult-to-Float Coal by Emulsion: A Review

Xin Wang [1], Huaigang Cheng [1,2,*] and Danjing Ding [1]

1 Institute of Resources and Environmental Engineering, Shanxi University, Taiyuan 030006, China; 15735175936@163.com (X.W.); 13485306952@163.com (D.D.)
2 College of Chemical Engineering, Qinghai University, Xining 810016, China
* Correspondence: chenghg@sxu.edu.cn; Tel.: +86-351-7018553

Abstract: Coal is expected to continue dominating the global energy landscape for a considerable period in the future. However, the depletion of high-quality coal resources and the increasing proportion of difficult-to-float coals are exacerbating environmental issues and leading to significant waste of carbon resources, making the clean and efficient utilization of such coals imperative. Enhancing the quality of coal through flotation is a prerequisite for the resource utilization of coal. Difficult-to-float coal, characterized by high hydrophilicity, complex pore structures, and fine particle size, poses challenges for efficient flotation using conventional collectors. Emulsions, owing to their exceptional surface and interfacial regulation capabilities and environmental adaptability, have been employed as flotation collectors for various minerals and have garnered significant attention in recent years for their application in the flotation of difficult-to-float coals. In the pursuit of green and cost-effective flotation technologies for such coals, this paper systematically reviews the causes of poor floatability in difficult-to-float coals and their latest research progress by emulsion flotation. It summarizes the impact of emulsion types and preparation methods on their properties and application areas, with a particular focus on the key mechanisms by which emulsion collectors enhance the flotation of difficult-to-float coals, including surface charge regulation, surface hydrophobicity modification, and interfacial tension control. Finally, this paper outlines future research directions on emulsion flotation, which will likely focus on the precise control of emulsion structure and size, the targeted separation of organic components by emulsion collectors under complex conditions, the development of low-cost and highly biocompatible synthetic reagents, and the development of efficient emulsion storage and transportation equipment.

Keywords: difficult-to-float coal; emulsion; flotation; mechanism

1. Introduction

Coal is the most significant primary energy source and an essential raw material in industrial production [1–3], and it is expected to continue to dominate the global energy structure for a considerable period in the future [4,5]. However, with the ongoing extraction of coal resources, the reserves of high-quality coal have sharply declined, while the proportion of difficult-to-float coal has steadily increased. The long-term accumulation of difficult-to-float coal can lead to numerous environmental issues and a significant waste of carbon resources [6–8]. Therefore, the large-scale, clean, and efficient utilization of difficult-to-float coal is an issue of growing concern.

Difficult-to-float coal, including low-rank coal, oxidized coal, and fine-grained coal, accounts for more than 70% of the proven coal reserves [9–11]. Low-rank coal, characterized by its low degree of coalification, is widely distributed across the globe [12]. Additionally, issues related to the oxidation of raw coal due to improper mining, transportation, or storage have not been effectively addressed [13]. The increasing intensity of mechanized mining has also led to a rising proportion of fine-grained coal (with a particle size smaller than 0.5 mm) within coal resources [14]. Due to its low calorific value, high ash content, and high volatile matter, the direct combustion of difficult-to-float coal is challenging. Therefore, to maximize the utilization value of coal resources and minimize environmental pollution, the coal industry is urgently seeking to achieve the clean and efficient utilization of difficult-to-float coal.

Froth flotation is the most economical and widely used technology for coal beneficiation and upgrading. The principle behind this technique is based on the difference in hydrophobicity between the organic carbon in coal and gangue minerals, which enables their separation. When the hydrophobicity of fine coal particles exceeds that of gangue particles, the fine coal particles efficiently attach to the surfaces of air bubbles, rising with them to the top of the slurry, while the gangue particles remain at the bottom, thereby achieving separation [15–17]. High-quality coal, with a high organic carbon content and strong hydrophobicity, can be easily separated from other hydrophilic minerals. However, the flotation efficiency of difficult-to-float coals, such as low-rank coal, is significantly reduced due to their high content of surface oxygen-containing functional groups and hydrophilic minerals, which hinder the adsorption and spreading of traditional hydrocarbon oil collectors [18]. Additionally, hydrocarbon oils are flammable, volatile, malodorous, and poorly biodegradable [19], posing risks of explosion and environmental contamination during storage and transportation.

To achieve the safe, clean, and efficient flotation of coal while reducing the use of hydrocarbon oils, the development of novel collectors with good dispersibility and strong stability is crucial [20–22]. Emulsions, which are multiphase systems composed of water, oil, and emulsifiers [23], possess a large specific surface area and strong interfacial regulation capabilities. The adsorption of suitable emulsifiers at the oil/water interface not only reduces interfacial tension and system energy, thereby promoting emulsion stability [24,25], but also enables the directional regulation of the contact interface between coal particles and the emulsion, enhancing the floatability of coal particles [26,27]. Furthermore, emulsion collectors with good biocompatibility not only exhibit high flotation efficiency for difficult-to-float coal but also offer excellent environmental friendliness. This aligns with the trend towards green chemistry and demonstrates broad application prospects in the efficient utilization of coal resources.

Researchers have extensively studied the application of emulsions in mineral flotation. Sahasrabudhe et al. were the first to apply high internal phase water-in-oil emulsions in the flotation of chalcopyrite, achieving the rapid and efficient recovery of chalcopyrite [28]. Suyantaraa et al. investigated the effect of emulsified kerosene on the flotation of molybdenite and chalcopyrite in simulated seawater, with flotation tests showing that the effective separation of the two minerals could be achieved at pH = 9 [29]. Li et al. used octyl hydroxamic acid as an emulsifier to emulsify kerosene, finding that the emulsified kerosene droplets were smaller and more uniform, allowing better dispersion on the surface of malachite in the flotation system, thereby improving its flotation recovery rate [30]. Additionally, the recovery of carbon resources using emulsion collectors has become a research hotspot in recent years. Sun et al. studied the differences in the flotation performance of fine flake graphite before and after kerosene emulsification, revealing that emulsified kerosene exhibited significantly better flotation performance than raw kerosene [31]. Nie et al. discovered that emulsifying waste cooking oil could effectively recover carbon from waste coal ash, achieving the goal of "waste treating waste" [32]. Compared to traditional oily collectors, emulsion collectors offer better dispersion and interfacial activity, not only enhancing the collection efficiency and selectivity of minerals but also reducing the environmental harm caused by the excessive use of oily collectors, thereby providing excellent economic and environmental benefits. Therefore, replacing traditional oils with emulsions as mineral flotation collectors is both necessary and feasible. Different methods can be used to prepare various types of emulsions, and the type of emulsion is closely related to its properties. Systematically elucidating the correlation between emulsion preparation methods, types, and properties is of great theoretical significance in improving the application of emulsions in mineral flotation, particularly for the flotation of difficult-to-float minerals such as low-rank coal.

In this paper, recent research findings are reviewed to summarize the types of emulsions, their preparation methods, and the progress made in the application of emulsion collectors for the beneficiation of difficult-to-float coal. Based on this, the primary mechanisms by which emulsion collectors enhance the flotation of difficult-to-float coal are discussed. Finally, future research trends on emulsion flotation are projected, with the aim of providing valuable insights and references for the research and application of emulsion flotation technology.

2. Methodology

This review aims to reveal the research status in the field of emulsion flotation for difficult-to-float coal through a comprehensive investigation and analysis of relevant literature. The main mechanisms by which emulsions enhance the flotation of difficult-to-float coal are identified, along with the challenges in this area. Furthermore, the review provides an outlook on future research directions, with the goal of offering clearer guidance and reference points for advancing emulsion flotation fordifficult-to-float coal.

In this review, the topic of emulsion flotation for difficult-to-float coal was first identified, followed by a thorough search and analysis of relevant research articles and published reviews. The selection of literature was conducted using databases such as Web of Science and Elsevier, based on the following keywords: (1) Emulsion flotation, (2) Low-rank coal flotation, (3) Oxidized coal flotation, (4) Fine-grained coal, (5) Coal flotation, and (6) Emulsifier flotation. These articles were then manually screened by reading the titles and abstracts. Approximately 300 articles closely related to the topic were selected, and further reading and evaluation were conducted based on specific subtopics. Ultimately, 85 articles were selected for comprehensive analysis and summary, which defined the themes and direction discussed in this review (Figure 1).

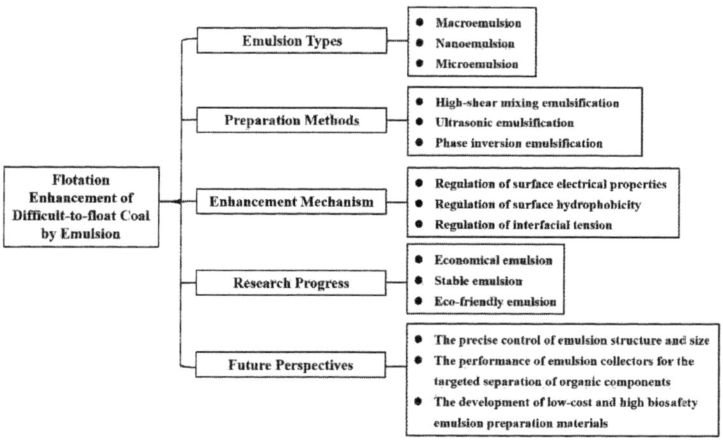

Figure 1. Topics discussed in this review.

3. The Difficult Flotation Mechanism of Difficult-to-Float Coal

The abundance of surface oxygen-containing functional groups is one of the primary reasons for the difficulty in efficiently floating difficult-to-float coal. On one hand, low-rank coal, which has undergone a lower degree of coalification, contains significantly higher amounts of oxygen-containing functional groups compared to medium-rank coals and high-rank coals. Moreover, during mining, storage, and transportation, coal is highly susceptible to oxidation due to exposure to natural factors such as wind, sunlight, and rainfall, further increasing the content of oxygen-containing functional groups on its surface [33]. The rich presence of these functional groups significantly enhances the hydrophilicity of low-rank/oxidized coal, leading to the formation of a dense hydration film on the surface, which hinders the effective attachment of air bubbles (as illustrated in Figure 2). Chen et al. [34] oxidized ultra-low ash coal samples with varying concentrations of hydrogen peroxide to prepare coals with different degrees of oxidation and found that the content of oxygen-containing functional groups such as carboxyl, hydroxyl, carbonyl, and ether groups increased significantly with the degree of oxidation. Multiple studies have shown that ether (R-O), carboxyl (-COOH), and hydroxyl (-OH) groups are the main contributors to the enhanced wettability of the coal surface. Li et al. found that as the content of oxygen-containing functional groups (hydroxyl) on the coal surface increased, the thickness of the water film on the coal surface increased, the arrangement of water molecules within the film became denser, their mobility decreased, and the number of hydrogen bonds between interfacial water molecules and surface hydroxyl groups increased [35]. These factors make it difficult for non-polar hydrocarbon oils to effectively spread and adsorb on the surface of low-rank/oxidized coal, thereby significantly increasing the consumption of hydrocarbon oils during the flotation process.

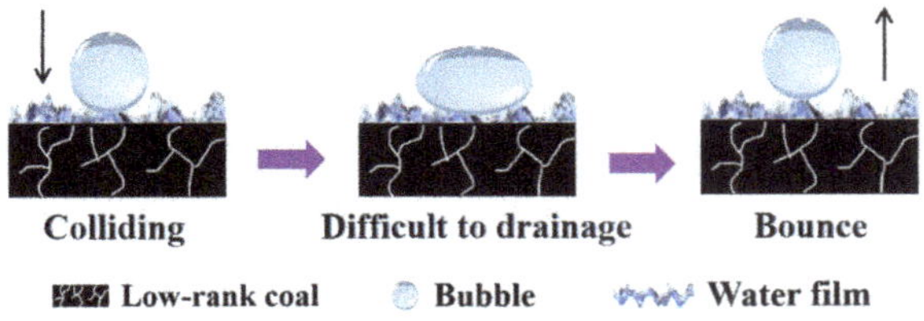

Figure 2. Schematic of water film hindering the adsorption of bubbles on the surface of low-rank coal (The direction of the arrows indicates the motion of the bubbles.).

High porosity is another factor contributing to the difficulty in efficiently floating difficult-to-float coal. Difficult-to-float coals, such as low-rank coal and oxidized coal, typically possess rough, porous surfaces, primarily in the form of micropores. Additionally, the collapse of some pores on the coal surface after oxidation can lead to the formation of more micropores. The porous structure facilitates the formation of a thick and stable hydration film on the coal surface. Shen et al. used molecular dynamics simulations to verify the role of pore spaces in promoting the formation of hydration layers, where water molecules above the pores are attracted by the water present on the coal surface and rapidly fill the pores [36]. In turn, the pores and cracks on the coal particle surface restrict the movement of water molecules. Due to the thick and stable hydration film, air bubbles are unable to penetrate the hydration layer to interact with the coal surface. As illustrated in Figure 3, the abundance of pores on the coal surface leads to increased collector consumption during the flotation process. The greater the number of pores, the larger the specific surface area, and consequently, the more collector is adsorbed, requiring a higher dosage of collector to achieve effective flotation.

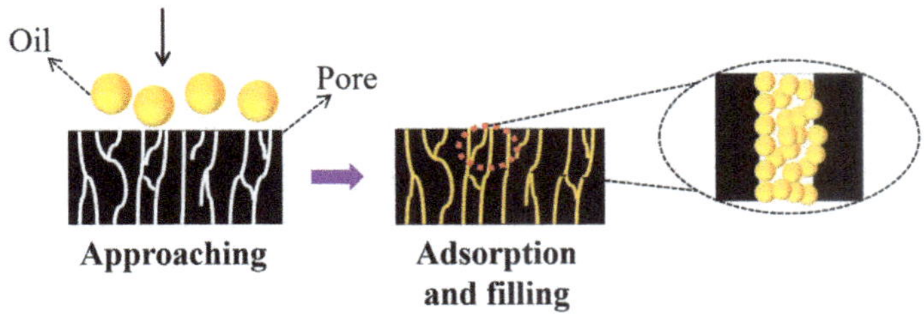

Figure 3. Schematic of adsorption and filling of oil collectors in low-rank coal pores.

The presence of hydrophilic gangue minerals is another significant factor contributing to the difficulty in efficiently floating difficult-to-float coal. Coal typically contains hydrophilic impurity minerals such as kaolinite, boehmite, and quartz (Figure 4). Due to the deterioration of geological conditions and the advancement of mechanized coal mining, the reserves of high-quality coal resources have gradually decreased, while the proportion of low-grade coal resources has increased. Low-grade coal resources often contain a large amount of gangue minerals. During coal mining and flotation processes, as the particle size decreases, these hydrophilic minerals form a hydrophilic layer between the coal particles and air bubbles, reducing the collision probability between the organic components of coal and the bubbles, thereby further increasing the difficulty of flotation.

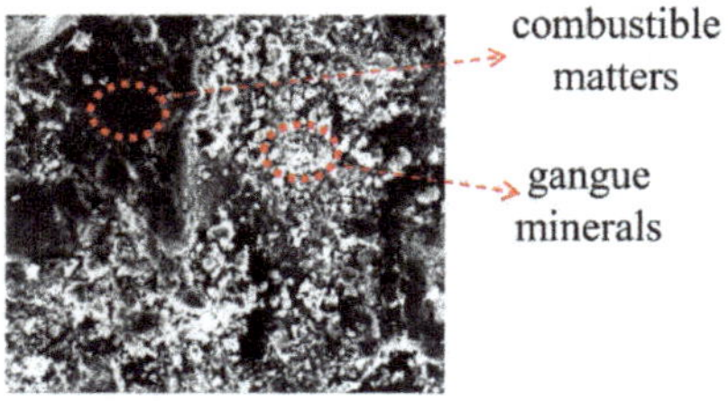

Figure 4. SEM images of the occurrence forms of gangue minerals in difficult-to-float coal.

Additionally, due to natural weathering and mechanical comminution, the particle size of coal continues to decrease [37]. In some coal preparation plants, the content of ultrafine coal particles (<0.074 mm) in fine coal can reach up to 90%, resulting in deteriorating floatability of the coal, a trend that is expected to worsen further, making the separation of fine coal increasingly challenging [38]. The fineness of coal particles represents a larger surface area and higher collector consumption, which can lead to increased froth entrainment during the flotation process, thereby affecting the quality of the clean coal. Therefore, the fineness of coal particles poses a significant challenge in the coal flotation separation process.

4. Fundamental Characteristics of Emulsions

4.1. Types and Characteristics of Emulsions

Emulsions can be classified into macroemulsions, nanoemulsions, and microemulsions based on droplet size [39], with their primary differences summarized in Table 1. The droplet size of macroemulsions typically exceeds 1 μm, while both nanoemulsions and microemulsions have droplet sizes on the nanometer scale. Therefore, the type of emulsion cannot be determined solely by droplet size [40,41]. Both macroemulsions and nanoemulsions are thermodynamically unstable systems, with the rate of phase separation depending on the energy barrier between the emulsion and the separated state. However, their physicochemical or thermodynamic properties do not exhibit significant differences. Nevertheless, nanoemulsions generally exhibit better stability than macroemulsions, including enhanced spatial stability against coalescence/flocculation and structural stability against Ostwald ripening.

The droplet size of microemulsions is also on the nanometer scale; however, compared to nanoemulsions, microemulsions have a narrower droplet size distribution and more uniform size, making them both thermodynamically and kinetically stable systems [42]. Unlike in macroemulsions and nanoemulsions, where surfactant molecules adsorb at the oil droplet interface, the particles in microemulsions are more akin to oil molecules that either coalesce as a distinct core or dissolve within micelles or between surfactant tails [43]. This characteristic imparts microemulsions with high thermodynamic stability while also making them highly sensitive to dilution or temperature changes.

Table 1. Comparison of properties of macroeumlsion, nanoemulsion, and microemulsion.

Item	Macroemulsion	Nanoemulsion	Microemulsion
Definition	A kinetically unstable multiphase system with a micron-sized dispersed phase within a continuous phase is formed under the influence of emulsifiers and external energy input.	A kinetically stable multiphase system with a nanoscale dispersed phase within a continuous phase is formed under the influence of emulsifiers and external energy input.	A thermodynamically stable multiphase system with a nanoscale dispersed phase within a continuous phase is formed under the influence of an emulsifier and changes in physicochemical conditions.
Particle size	1–100 μm	20–500 nm	10–100 nm
Stability	Thermodynamically unstable; weakly kinetically stable	Thermodynamically unstable; kinetically stable	Thermodynamically stable; kinetically stable
Polydispersity	High	Low	Very low
Optical properties	Milk-like	Translucent, transparent	Transparent

The type of emulsion has a significant impact on its characteristics, which, in turn, determine its areas of application. Macroemulsions, with their larger droplet sizes and lower stability, are commonly used in industrial fields such as mineral processing and oil-field extraction. Nanoemulsions and microemulsions, due to their extremely small droplet sizes and superior biocompatibility, are frequently employed in the fine chemicals sector, including pharmaceuticals, food, and cosmetics. Selecting the appropriate type of emulsion based on the application requirements is crucial for achieving the desired functionality.

4.2. Preparation Methods and Characteristics of Emulsions

The preparation methods of emulsions can be categorized into high-energy and low-energy methods. High-energy methods primarily include high-shear mixing and ultrasonic emulsification, which require the use of mechanical equipment with strong emulsifying capabilities, such as high-shear mixers, high-pressure homogenizers, and ultrasonic generators [44–46]. Low-energy methods mainly involve phase inversion emulsification, which utilizes interfacial energy within the system to prepare emulsions [47]. As shown in Table 2, the principles and applicable systems of different preparation methods vary, leading to differences in performance. Selecting the appropriate preparation method based on the intended application of the emulsion is crucial for enhancing the efficiency and cost-effectiveness of emulsion preparation.

Table 2. Comparison of emulsion preparation methods.

Method	Mechanism	Type	Field	Advantage (A)/ Disadvantage (D)	Reference
High-shear mixing emulsification	Under the action of emulsifiers, stable emulsions are formed by mixing the oil phase and water phase under high-shear conditions using high-shear equipment, such as homogenizers or high-speed shear mixers.	Macroemulsions	The requirements for particle size and stability are not critical, such as mineral processing and oil extraction.	A: High efficiency; good particle size control; wide applicability, mature preparation equipment D: High energy consumption; low stability	[48]
Ultrasonic emulsification	Under the action of emulsifiers, stable emulsions are formed by breaking up droplets through the cavitation effect generated by high-frequency ultrasonic vibration.	Macroemulsions and nanoemulsions	The requirements for high stability and uniformity, such as the food and pharmaceutical industries.	A: High efficiency; good particle size control; easy to operate; suitable for sensitive substances. D: High energy consumption; low efficiency; low stability.	[49]

Table 2. *Cont.*

Method	Mechanism	Type	Field	Advantage (A)/ Disadvantage (D)	Reference
Phase inversion emulsification	By altering the composition or physicochemical conditions, the redistribution of surfactants within the emulsion system is induced, leading to changes in the mutual solubility and interfacial tension between the oil and water phases, thereby forming a stable emulsion.	Microemulsions	The stringent requirements for stability, dispersibility, and particle size control, such as precision drug delivery and high-performance cosmetics.	A: Low energy consumption; precise particle size control; high stability D: Strict preparation conditions; low efficiency.	[47,50]

4.2.1. High-Shear Mixing Emulsification

High-shear mixing emulsification applies intense shear forces, typically using a high-speed rotor or homogenizer, to disperse immiscible liquids into fine droplets within the continuous phase, resulting in a stable emulsion with a controlled droplet size distribution. This method is currently primarily used for the preparation of macroemulsions. Shear mixing (as illustrated in Figure 5) influences the dispersibility and stability of emulsions by altering interfacial tension, rheological properties, and droplet distribution [48,51]. Studies have found that appropriate shear mixing time and rate can effectively enhance the stability of emulsions, while excessively high shear rate and prolonged shear mixing time can accelerate droplet coalescence through Brownian flocculation, leading to emulsion destabilization [52]. The viscosity of an emulsion is closely related to the shear rate, and researchers have utilized various models to investigate the relationship between shear rate and emulsion viscosity [53]. It has been observed that the relative viscosity of the emulsion decreases with increasing shear rate, and the viscosity of the emulsion is more sensitive to changes in shear rate under low-temperature conditions [54].

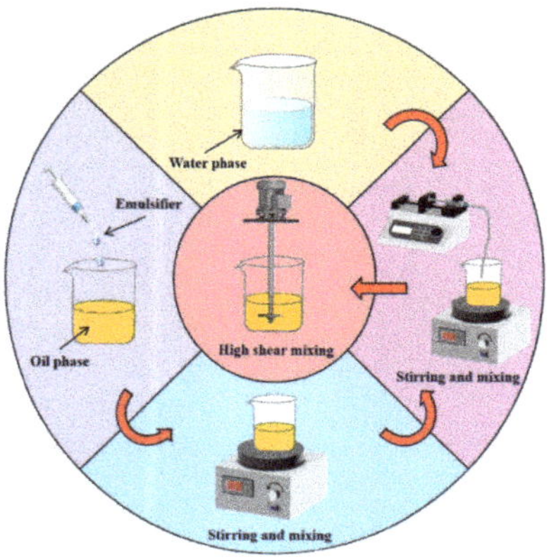

Figure 5. Process of emulsion preparation by high shear method.

4.2.2. Ultrasonic Emulsification

Ultrasound emulsification primarily disperses liquids into fine droplets within the continuous phase by applying ultrasonic waves, generating localized high-energy cavitation, which leads to the formation of a stable emulsion with a narrow droplet size distribution. Ultrasonic emulsification, due to its high power output, is commonly used to prepare emulsions with smaller droplet sizes, such as microemulsions and nanoemulsions. Due to its strong dispersing effect, ultrasound emulsification is commonly used for the preparation of macroemulsions with smaller particle sizes or nanoemulsions [55,56]. The ultrasonic preparation of emulsions is primarily achieved through the phenomenon of acoustic cavitation (as illustrated in Figure 4), which involves processes of bubble nucleation, growth, oscillation, and collapse. Ultrasonic cavitation involves the processes of bubble nucleation, growth, oscillation, and collapse. Figure 6 illustrates the fundamental process of ultrasonic cavitation by depicting schematic diagrams and high-speed camera recordings of the changing states of bubbles in water under the influence of ultrasound. The efficiency of cavitation during the ultrasonic process significantly impacts the emulsification, and various factors, such as ultrasonic frequency, power, temperature, and the surface tension or viscosity of the medium, can influence cavitation efficiency [57,58]. In the low-frequency range, particularly at 20 kHz, ultrasound is conducive to producing stable emulsions [59]. Additionally, increasing the ultrasonic power enhances the sonochemical effects [60]. As the ultrasonic power increases, both the number and average size of bubbles rise, resulting in stronger shear forces during bubble collapse, which reduces oil droplet size and improves emulsion stability. Further studies have shown that the impact of ultrasonic amplitude on droplet size reduction is greater than that of ultrasonic duration, while increasing the processing volume can reduce the system's energy density, thereby leading to smaller droplet sizes [49].

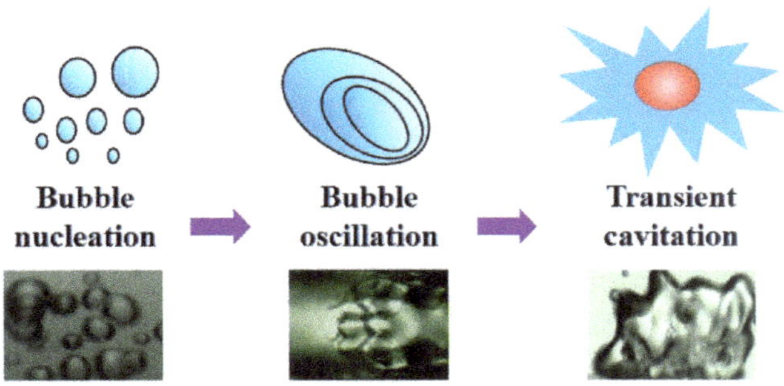

Figure 6. Schematic of ultrasonic cavitation.

4.2.3. Phase Inversion Emulsification

Phase inversion emulsification is a method of forming emulsions by altering the composition or physicochemical conditions, such as temperature or concentration, causing the roles of the dispersed phase and continuous phase to reverse. This process spontaneously results in microemulsions with fine droplet sizes and high stability. The phase inversion emulsification process involves both phase inversion temperature (PIT) and phase inversion composition (PIC), and it is primarily used for the preparation of nanoscale emulsions with smaller droplet sizes [50]. The PIT method is suitable for surfactants that transition from water-soluble (hydrophilic) to oil-soluble (hydrophobic) as the temperature increases. Below the so-called hydrophilic/lipophilic balance (HLB) temperature (THLB), the surfactant chains are highly hydrated, leading to their self-assembly at interfaces with positive curvature, forming O/W droplets. When the temperature exceeds THLB, the curvature of the surface interface becomes negative, resulting in the formation of W/O nanodroplets (Figure 7a). As the temperature approaches THLB, the spontaneous curvature approaches zero. At this point, a bicontinuous microemulsion containing significant amounts of both the aqueous and oil phases coexists with excess aqueous and oil phases, representing the intermediate state of the phase transition [61–63]. In contrast, the PIC method is conducted under isothermal conditions (Figure 7b) [64,65]. Typically, water is added to a mixture of oil and surfactants, inducing changes in the curvature of the surfactant layer, which leads to the production of O/W nanoemulsions. Compared to the PIT method, which requires strict control of conditions, the PIC method involves a simple dilution process, making it suitable for scale-up production but less suitable for high-temperature environments. Additionally, the PIC method is flexible and applicable to both nonionic and ionic surfactants. Similar to the PIT method, the PIC method also exhibits an intermediate state with zero-curvature structures, such as bicontinuous microemulsions or flexible lamellar liquid crystals, as the composition continues to change [66].

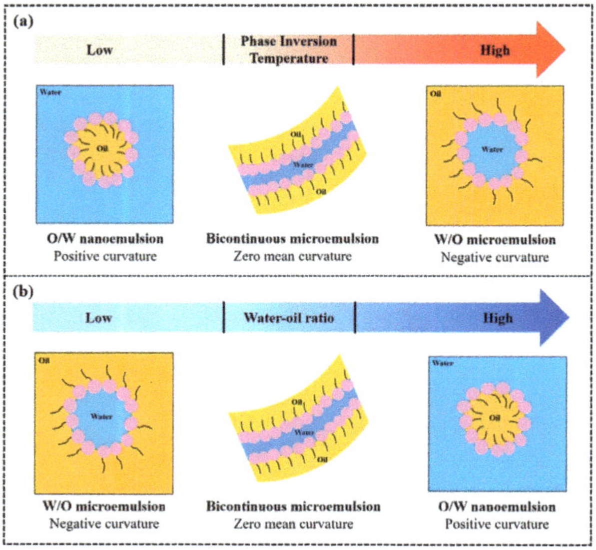

Figure 7. Schematics of the structural and curvature changes during (**a**) PIT and (**b**) PIC.

Currently, emulsion collectors are primarily composed of macroemulsions, which are mainly prepared using high-shear mixing emulsification. In recent years, the ultrasonic emulsification method for preparing microemulsions has also attracted increasing attention, while phase inversion emulsification is mainly employed in fields such as food and pharmaceuticals, where strict control over emulsion droplet size is required. In recent years, ultrasonic emulsification has attracted increasing attention for the preparation of coarse emulsions or nanoemulsions with high uniformity and small particle sizes. In contrast, phase inversion emulsification is primarily used in fields that demand strict control over emulsion particle size, such as high-precision drug delivery and high-performance cosmetics. Among these methods, high-shear mixing emulsification remains the main technology for the preparation of emulsion collectors due to its mature processes and flexible environmental requirements. However, the outstanding performance of microemulsions is gaining more attention, and with continuous advancements in preparation techniques, they are expected to have broader application prospects in the future.

5. Research Advances on Emulsion Flotation of Difficult-to-Float Coal

Non-polar hydrocarbon oils are the most commonly used collectors in coal flotation [67]. For difficult-to-float coal, due to the abundance of surface functional groups, high content of hydrophilic gangue minerals, fine particle size, and susceptibility to entrainment, achieving efficient and economical flotation using traditional hydrocarbon oil collectors is challenging [23]. In recent years, emulsions have attracted significant attention in the field of mineral flotation due to their excellent interfacial regulation capabilities and favorable economic performance.

High internal phase water-in-oil emulsions (HIP W/O) collectors replace a significant portion of the oil phase with an internal water phase, thereby greatly reducing oil consumption while maintaining the collector characteristics of the oil phase. It offers both efficiency and cost-effectiveness. van Netten et al. [68] were the first to successfully apply water-in-oil (W/O) emulsions for the beneficiation of fine coal. The high HIP W/O emulsion achieved a high combustible recovery rate with only 1/10 of the hydrocarbon oil used in traditional processes. LU et al. [69] successfully prepared a HIP W/O emulsion using kerosene as the external phase, achieving a clean coal recovery rate of 86.43% with an oil consumption of approximately 1/5 of that of pure kerosene. Furthermore, compared to pure kerosene, the viscosity of the high internal phase emulsion is significantly increased, which allows for more efficient adhesion following collisions between the emulsion and coal particles, thereby enhancing flotation efficiency [70,71]. Mashaba et al. [11] used a HIP W/O emulsion with a 95% internal water phase for the flotation of weathered South African coal. Compared to kerosene, the ash content of the clean coal was reduced by 5.71%, and the calorific value increased by 2.19 MJ/kg.

Microemulsions are thermodynamically stable systems and, compared to conventional emulsion collectors, offer advantages such as higher stability, smaller droplet size, higher surface activity, and better dispersibility. The preparation of microemulsions primarily employs phase inversion emulsification, which requires strict control of preparation conditions. With ongoing advancements in the theory and technology of microemulsion preparation, microemulsions have recently garnered widespread attention as collectors for the flotation of difficult-to-float coal. Xue et al. [72] prepared microemulsions of diesel using ethanol ethoxylate (AEO-3) as an emulsifier and found that the addition of AEO-3 significantly reduced surface tension, produced smaller emulsion droplets, and could adhere to coal particles to enhance their hydrophobicity. Molecular dynamics simulations indicated that the addition of AEO-3 strengthens the electrostatic interaction between collector (diesel) molecules and coal particles, reducing their mean square displacement. Moreover, microemulsions can increase the contact angle between particles and bubbles, accelerating the formation of stable mineral flocs, which is a key factor in the faster flotation process. Du et al. [73] conducted flotation experiments on low-rank coal fines using microemulsions prepared from a composite emulsifier and diesel, achieving optimal flotation performance at an emulsifier-to-diesel ratio of 1:2.

The "waste-to-resource" approach significantly reduces carbon emissions by converting waste materials into resources, thereby decreasing the demand for new raw materials. Under the "Dual Carbon" strategy, the significance of the "waste-to-resource" concept is particularly prominent. Many researchers have begun exploring the possibility of using waste oil to prepare efficient emulsion collectors. Cheng et al. [26,74] developed a novel oil-in-water (O/W) emulsion collector using waste cooking oil as the raw material. Compared to diesel and waste cooking oil collectors, the waste frying oil emulsion collector (WFOCE) demonstrated superior flotation performance, with a 70% reduction in the required dosage after emulsification, offering substantial economic and environmental benefits.

Furthermore, with the growing prominence of global environmental issues, "Green mineral Processing" has emerged as a key trend in the industry. The use of eco-friendly collectors not only enables efficient flotation but also significantly reduces environmental impact, achieving a balance between resource utilization and environmental protection. Based on this, researchers have focused on the development and utilization of renewable and highly biocompatible collectors. Gong et al. [75] prepared emulsions using renewable oil synthesized via the Fischer-Tropsch process as the oil phase and five different emulsifiers, respectively. The results indicated that the O/W emulsion prepared with Tween 80 as the emulsifier exhibited greater stability and better flotation performance. Similarly, Li et al. [76] emulsified kerosene using the biocompatible surfactant alkyl polyglycoside (APG) to prepare a W/O emulsion and optimized the preparation conditions using phase diagrams. The micro-flotation results of low-rank coal showed that the combustible recovery increased by 31.4%, while kerosene consumption decreased by 57.8%. The emulsifier molecules acted as a "bridging agent" between the oxygen-containing groups on the coal surface and the kerosene in the emulsion, significantly enhancing the adsorption efficiency of the emulsion on the surface of low-rank coal.

Oily collectors have been widely applied in the industrial flotation of various hydrophobic minerals, with low cost and easy availability. For the flotation of more challenging hydrophilic minerals, such as difficult-to-float coal, their large droplet size and weak adhesion limit their dispersion efficiency, requiring higher dosages to achieve the desired effect. In contrast, emulsion collectors feature smaller droplet sizes and higher interfacial activity, which significantly enhances their probability of contact and adsorption efficiency with the minerals. Moreover, the flexible structure and composition of emulsions can be directionally adjusted according to the surface properties of the floated minerals, enabling more selective mineral recovery. Additionally, their environmental friendliness aligns better with the development trends of modern industries. Further reducing the production cost is crucial for promoting their industrial application. The research status of emulsion flotation for difficult-to-float coal is presented in Table 3.

Table 3. Research status on flotation of difficult-to-float coal by emulsion.

Direction	Type	Challenge	Reference
Low cost	HIP-macroemulsion; HIP-nanoemulsion	Emulsion stability requires improvement	[11,69–71]
High performance	Microemulsion	Preparation process has low efficiency and high cost	[72,73]
Environmental friendliness	Macroemulsion; Nanoemulsion; Microemulsion	The raw materials are scarce, and the preparation process is complex.	[27,74]

6. Enhancement Flotation Mechanism of Difficult-to-Float Coal by Emulsion

The primary enhancement mechanism of emulsion-enhanced flotation for difficult-to-float coal is based on the regulation of interfacial properties through the action of emulsifiers. This method modifies only the contact interfaces or the surface properties of the coal, without affecting its internal structure.

The regulation of interfacial properties refers to the combination of emulsions and coal particles through van der Waals forces, hydrogen bonding, π-π interactions, and other intermolecular forces. This process changes the interfacial tension, surface charge, and hydrophobicity at the contact interface or on the coal surface, thereby achieving efficient coal flotation. The underlying mechanism is illustrated in Figure 6. Figure 8 illustrates the possible mechanism of action for enhancing the flotation of difficult-to-float coal using mainstream macroemulsion collectors. The amphiphilic nature of emulsifier molecules can alter the properties of the interface between the internal and external phases of the emulsion, promoting emulsion formation and enhancing its stability. Simultaneously, this amphiphilic nature allows for the directional regulation of the contact interface between the coal particles and the emulsions, making it favorable for coal flotation [22,77,78]. Emulsion collection leverages this property to enhance the separation and upgrading of difficult-to-float coal during the flotation process.

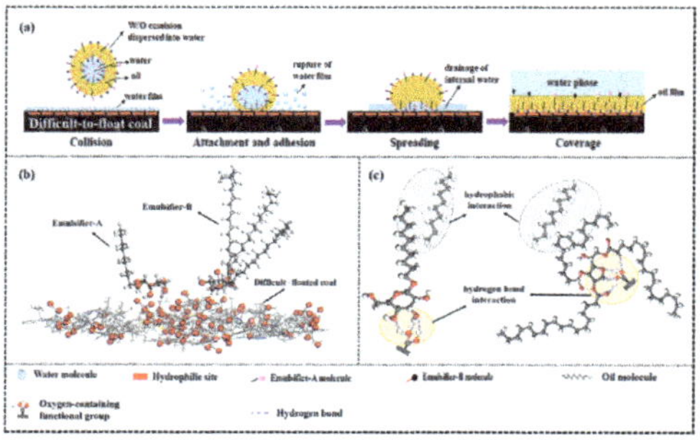

Figure 8. (**a**) Interaction between macroemulsion and difficult-floated coal surface. (**b**) Adsorption behavior of macroemulsion at difficult-floated coal surface. (**c**) Interfacial molecular interaction of emulsifiers at difficult-floated coal surface.

6.1. Regulation of Surface Electrical Properties

The surface electrical properties of difficult-to-float coal significantly influence its dispersion, aggregation, hydrophobicity, and the adsorption of reagents in the slurry. These properties are crucial factors determining whether the coal can be efficiently floated. When emulsions are adsorbed onto the surface of difficult-to-float coal, they can regulate electrical properties of the coal surface, altering them towards a more easily separable state. Xing et al. [22] applied a formulated emulsion for the recovery of combustible components in low-rank coal and found that the surface potential of the raw coal increased from −50.41 mV to −29.58 mV. This effectively reduced the electrostatic repulsion between coal particles and air bubbles, enhancing the adhesion between them and thereby improving the flotation efficiency of combustible components. In another study by Du, it was found that after adding SPS microemulsions, the Zeta potential of low-rank coal slurry significantly decreased. At pH = 7, the absolute value of the Zeta potential was 8.67 mV higher than that of the raw coal. This is because sodium dodecyl sulfate (SDS), an anionic emulsifier in the SPS emulsion, carries a negative charge in water [73]. The increased electronegativity enhances the repulsive forces between coal particles, allowing finer coal particles to disperse more effectively in the slurry, come into contact with the collectors, and facilitate flotation [79].

6.2. Regulation of Surface Hydrophobicity

The hydrophobicity of mineral surfaces is also a key factor in determining whether they can be efficiently floated. It is a principle utilized in the flotation decontamination devices for many industrial solid wastes, such as the removal of impurity carbon from carbide slag [80]. Difficult-to-float coal, due to its surface being rich in oxygen-containing functional groups and a high content of hydrophilic impurity minerals, exhibits strong hydrophilicity, which is unfavorable for flotation. Emulsion collectors can significantly improve the hydrophobicity of difficult-to-float coal surfaces, enhance the adhesion between bubbles and difficult-to-float coal, and increase the flotation efficiency. This improvement is due to the amphiphilic nature of the emulsion; the hydrophilic head groups of the emulsifier adsorb onto the hydrophilic functional groups on the surface of the difficult-to-float coal, while the hydrophobic tail chains extend outward, enhancing the surface hydrophobicity of the difficult-to-float coal and making it easier to float and separate [81].

6.3. Regulation of Interfacial Tension

During the flotation process, the interfacial tension at the contact interface between the collector and coal particles largely determines the adsorption and spreading efficiency of the collector molecules on the coal surface. The lower the interfacial tension, the higher the adsorption efficiency of the collector onto the coal particles. Studies have shown that emulsion regulation can significantly reduce the free energy and interfacial tension at the contact interface, decrease the spreading energy of the emulsion on the coal surface, and lower the potential energy of interactions between them. This enhances the adsorption strength of the emulsion on the coal surface and improves flotation efficiency [69]. Zhang et al. added laurylamine dipropylenediamine (LDD) to the kerosene flotation system for oxidized coal and measured the surface tension of the system before and after the addition using the Wilhelmy plate method. It was found that the addition of LDD effectively reduced the interfacial tension of the system from 72 mN/m to 41 mN/m, thereby significantly enhancing the adsorption efficiency of kerosene on the oxidized coal surface [82]. Ding et al. used the pendant drop method to study the effect of collector interfacial tension on the flotation of coal with different particle sizes and reached similar conclusions [83]. Utilizing this principle, emulsion collectors can significantly enhance the flotation efficiency of difficult-to-float coal.

Based on the above discussion, the enhancement of the flotation process for difficult-to-float coal by emulsions is primarily attributed to the emulsifiers' regulation of the coal particles' interfacial properties. The fundamental reason lies in the emulsifiers' ability to form chemical bonds such as hydrogen bonds and π-bonds with difficult-to-float coal, thereby promoting the adsorption of the emulsions onto the coal surface. In contrast, gangue minerals, which are typically composed of silicates, carbonates, or oxides, have surfaces dominated by polar groups, such as hydroxyl groups or oxygen ions, making them more hydrophilic than difficult-to-float coal. This strong hydrophilicity causes gangue minerals to resist emulsifier adsorption and remain in the aqueous phase during the flotation process.

7. Conclusions

The study of efficient flotation technologies for difficult-to-float coal is driven by the need for strategic technological reserves and capacity reserves in global energy strategies. Emulsion collectors have demonstrated a significant enhancing effect on the flotation of difficult-to-float coal, showcasing distinct technical advantages. After reviewing the articles and patents compiled in this study, the following conclusions were drawn.

(1) The low flotation efficiency of difficult-to-float coal using traditional hydrocarbon oil collectors is primarily due to its high content of oxygen-containing functional groups, high porosity, significant presence of hydrophilic gangue minerals, and fine particle size.
(2) Different types of emulsions require distinct preparation methods and possess varying structures and characteristics, which in turn determine their application areas. Macroemulsions are suitable for industrial applications where droplet size is not the most critical factor, while nanoemulsions and microemulsions receive more attention in fine chemical fields such as pharmaceuticals and food. With advancements in preparation theory and processes, microemulsions have garnered increasing attention from researchers, presenting a promising application potential in the flotation of difficult-to-float coal.
(3) In recent years, research on emulsion collectors has primarily focused on the following areas: the development and application of high internal phase emulsion collectors with good economic viability; the preparation and application of microemulsion collectors with enhanced stability; the development and application of waste oil emulsion collectors and eco-friendly emulsion collectors and the development of efficient emulsion storage and transportation equipment.
(4) The primary enhancement mechanism of flotation for difficult-to-float coal by emulsions is based on the regulation of interfacial properties, which includes the regulation of surface electrical properties, surface hydrophobicity, and interfacial tension. This method modifies only the contact interfaces or the surface properties of the coal, without affecting its internal structure.

8. Future Perspectives

Research on flotation enhancement of difficult-to-float coal by emulsion is actively underway. However, most studies remain at the experimental stage, with several issues yet to be addressed before practical application can be achieved. Based on the progress of theoretical research and the demands of practical applications, future research directions for flotation enhancement of difficult-to-float coal may include the following aspects.

(1) Research on the precise control of emulsion structure and size is crucial. The structure and size of emulsions are important factors affecting their stability and dispersibility. Developing cost-effective and efficient methods to prepare emulsions with appropriate structure and size is essential for enhancing flotation efficiency.
(2) Research on the performance of emulsion collectors for the targeted separation of organic components in complex environments is essential. Difficult-to-float coal has a diverse structure and complex composition, with organic components and inorganic minerals closely intergrown. The coexistence of multiple components in such a complex environment significantly affects flotation efficiency and the quality of clean coal. Enhancing the ability of emulsion collectors to selectively separate organic components in a multi-component environment will be a key focus of future research.

(3) The development of low-cost, highly biocompatible synthetic reagents is critical. The preparation cost and biocompatibility of emulsions are directly related to their potential application in the industrial flotation of difficult-to-float coal. Developing cost-effective and biocompatible emulsion-based synthetic reagents is of great significance for advancing the industrial application of emulsion flotation for difficult-to-float coal.
(4) The development of efficient storage and transportation equipment for emulsions is essential. Inappropriate storage and transport conditions can increase the likelihood of emulsion demulsification, disrupt emulsion structure, and negatively affect performance. To achieve the industrial application of emulsion collectors, efficient storage and transportation equipment are a critical factor.

In conclusion, future developments in emulsion flotation technology will focus on further refining the underlying theory, optimizing processes and equipment, and gradually applying these advancements to practical production. This will offer new approaches and technological pathways for the green and clean utilization of difficult-to-float coal.

Author Contributions: Writing—original draft preparation, X.W.; writing—review and editing, X.W.; supervision, H.C.; project administration, D.D.; funding acquisition, H.C. All authors have read and agreed to the published version of the manuscript.

Funding: This research was funded by the National Key Research and Development Plan of China (2022YFB4102102), Centrally Guided Local Science and Technology Development Fund Projects of Qinghai Province (2024ZY006), National Natural Science Foundation of China (U20A20149, 22478232), Shanxi University Interdisciplinary Construction Project (2023). We are grateful for their financial support and valuable guidance.

Data Availability Statement: Data are contained within the article.

Acknowledgments: Xin Wang would like to thank the Institute of Resources and Environmental Engineering at Shanxi University for making this work possible.

Conflicts of Interest: The authors declare no conflicts of interest.

References

1. Sivrikaya, O. Cleaning study of a low-rank lignite with DMS, Reichert spiral and flotation. *Fuel* **2014**, *119*, 252–258. [CrossRef]
2. Huang, G.; Xu, J.; Geng, P.; Li, J. Carrier Flotation of Low-Rank Coal with Polystyrene. *Minerals* **2020**, *10*, 452. [CrossRef]
3. Li, Y.; Xia, W.; Pan, L.; Tian, F.; Peng, Y.; Xie, G.; Li, Y. Flotation of Low-Rank Coal Using Sodium Oleate and Sodium Hexametaphosphate. *J. Clean. Prod.* **2020**, *261*, 121216. [CrossRef]
4. Gao, L.; Li, X.; Lyu, X.; Zhu, X. Advances and Perspectives of Green and Sustainable Flotation of Low-Rank/Oxidized Coal: A Review. *Energy Fuels* **2024**, *38*, 1566–1592. [CrossRef]
5. Lin, Q.; Mei, Y.; Huang, W.; Zhang, B.; Liu, K. Understanding the Role of Polyvinylpyrrolidone on Ultrafine Low-Rank Coal Flotation. *ACS Omega* **2022**, *7*, 10196–10204. [CrossRef]
6. Xu, M.; Zhou, Y.; Hao, Y.; Cao, Y.; Xing, Y.; Gui, X. Enhancing Flotation Performance of Low-Rank Coal Using Environment-Friendly Vegetable Oil. *Minerals* **2023**, *13*, 717. [CrossRef]
7. Liao, Y.; Yang, Z.; An, M.; Cao, Y.; Hao, X.; Song, X.; Ren, H.; Yang, A.; Chen, L. Spreading behavior of dodecane-oleic acid collector mixture in low-rank coal flotation. *Fuel* **2022**, *308*, 122071. [CrossRef]
8. Li, R.; Xie, Z.; Zhou, Y.; Wang, W.; Gui, X. Investigation of synergistic mechanism of magnesium ions and cationic surfactant in low-rank coal flotation. *J. Ind. Eng. Chem.* **2024**, *137*, 280–292. [CrossRef]
9. Xia, Y.; Xing, Y.; Gui, X. Oily collector pre-dispersion for enhanced surface adsorption during fine low-rank coal flotation. *J. Ind. Eng. Chem.* **2020**, *82*, 303–308. [CrossRef]
10. Wang, Y.; Li, Y.; Xia, W.; Peng, Y.; Xie, G. Adsorption of sodium oleate on coal surface with different oxidation degrees and its effect on flotation. *Colloids Surf. A Physicochem. Eng. Asp.* **2021**, *611*, 125801. [CrossRef]
11. Mashaba, P.M.; Bada, S.O. Flotation of weathered coal discards using a HIP W/O emulsion binder and kerosene. *J. South. Afr. Inst. Min. Metall.* **2022**, *122*, 437–442. [CrossRef] [PubMed]
12. Xia, Y.; Zhang, R.; Cao, Y.; Xing, Y.; Gui, X. Role of molecular simulation in understanding the mechanism of low-rank coal flotation: A review. *Fuel* **2020**, *262*, 116535. [CrossRef]
13. Vilaso-Cadre, J.E.; Ávila-Márquez, D.M.; Reyes-Dominguez, I.A.; Blanco-Flores, A.; Gutierrez-Castaneda, E.J. Coal flotation in a low-rank carbonaceous mineral using 3-phenyl-1-pro-panol as a collector reagent. *Fuel* **2021**, *304*, 121363. [CrossRef]
14. Zhao, X.; Tang, Y.; Zhao, B.; Wu, C.; Ding, Y. Collecting behaviors of high internal phase (HIP) emulsion in flotation of ultrafine high-ash content coal slime. *Int. J. Coal Prep. Util.* **2022**, *42*, 2635–2655. [CrossRef]

15. Lu, Y.; Li, E.; Cheng, H.; Wang, X.; Du, Z.; Cheng, F.; Miller, J.D. Effect of Oxygen Functional Groups on the Surface Properties and Flotation Response of Fine Coal, Comparison of Rank with Oxidation. *Int. J. Coal Prep. Util.* **2021**, *41*, 290–306. [CrossRef]
16. Liu, J.; Zhang, R.; Bao, X.; Hao, Y.; Gui, X.; Xing, Y. New insight into the role of the emulsified diesel droplet size in low rank coal flotation. *Fuel* **2023**, *338*, 127388. [CrossRef]
17. Hu, X.; Li, Y.; Li, W. Recycling of waste plastic combustion soot and diesel oil mixing to prepare a new collector to improve the performance of low-rank coal flotation. *Powder Technol.* **2024**, *439*, 119727. [CrossRef]
18. Xia, W.; Xie, G.; Peng, Y. Recent advances in beneficiation for low rank coals. *Powder Technol.* **2015**, *277*, 206–221. [CrossRef]
19. Chen, S.; Tang, L.; Tao, X.; He, H.; Chen, L.; Yang, Z. Enhancing flotation performance of low rank coal by improving its hydrophobicity and the property of oily bubbles using 2-ethylhexanol. *Int. J. Miner. Process.* **2017**, *167*, 61–67. [CrossRef]
20. Xia, Y.; Yang, Z.; Zhang, R.; Xing, Y.; Gui, X. Performance of used lubricating oil as flotation collector for the recovery of clean low-rank coal. *Fuel* **2019**, *239*, 717–725. [CrossRef]
21. Li, M.; Xia, Y.; Zhang, Y.; Ding, S.; Rong, G.; Cao, Y.; Xing, Y.; Gui, X. Mechanism of shale oil as an effective collector for oxidized coal flotation: From bubble–particle attachment and detachment point of view. *Fuel* **2019**, *255*, 115885. [CrossRef]
22. Xing, Y.; Gui, X.; Cao, Y.; Wang, D.; Zhang, H. Clean low-rank-coal purification technique combining cyclonic-static microbubble flotation column with collector emulsification. *J. Clean. Prod.* **2017**, *153*, 657–672. [CrossRef]
23. Duan, H.; Chang, H.H.; Gao, L.X.; Gai, K.D.; Yu, Y.X.; Lyu, X.J.; Zhu, X.N. Preparation of renewable collector from waste engine oil for clean flotation of fine coal with different oxidation degree. *Fuel* **2023**, *335*, 127003. [CrossRef]
24. Mao, L.; Dai, H.; Du, J.; Feng, X.; Ma, L.; Zhu, H.; Chen, H.; Wang, H.; Zhang, Y. Gelatin microgel-stabilized high internal phase emulsion for easy industrialization: Preparation, interfacial behavior and physical stability. *Innov. Food Sci. Emerg. Technol.* **2022**, *78*, 103011. [CrossRef]
25. Tan, H.; Sun, G.; Lin, W.; Mu, C.; Ngai, T. Gelatin Particle-Stabilized High Internal Phase Emulsions as Nutraceutical Containers. *ACS Appl. Mater. Interfaces* **2014**, *6*, 13977–13984. [CrossRef]
26. Cheng, G.; Zhang, M.; Lu, Y.; Zhang, H.; Von Lau, E. New insights for improving low-rank coal flotation performance via emulsified waste fried oil collector. *Fuel* **2024**, *357*, 129925. [CrossRef]
27. Cheng, G.; Zhang, M.; Lu, Y.; Zhang, Y.; Lin, B.; Von Lau, E. A novel method for the green utilization of waste fried oil. *Particuology* **2024**, *84*, 1–11. [CrossRef]
28. Sahasrabudhe, G.; DeIuliis, G.; Davy, J.; Galvin, K.P. Selective and ultrafast agglomeration of chalcopyrite by a water in oil emulsion binder. *Miner. Eng.* **2021**, *167*, 106900. [CrossRef]
29. Suyantara, G.P.W.; Hirajima, T.; Elmahdy, A.M.; Miki, H.; Sasaki, K. Effect of kerosene emulsion in $MgCl_2$ solution on the kinetics of bubble interactions with molybdenite and chalcopyrite. *Colloids Surf. A-Physicochem. Eng. Asp.* **2016**, *501*, 98–113. [CrossRef]
30. Li, Z.L.; Rao, F.; Lou, X.M.; Song, S.X.; López, A. Floc-Flotation of Malachite Fines with an Octyl Hydroxamate and Kerosene Mixture. *Minerals* **2019**, *9*, 301. [CrossRef]
31. Sun, K.K.; Qiu, Y.S.; Zhang, L.Y.; Liu, Q.P.; Mao, Z.F.; Qian, Y.P. Enhanced fine flake graphite flotation and reduced carbon emission by a novel water-in-oil kerosene emulsion. *Colloids Surf. A-Physicochem. Eng. Asp.* **2022**, *650*, 129603. [CrossRef]
32. Nie, C.C.; Shi, S.X.; Wang, J.X.; He, Q.M.; Qi, X.F.; Xue, P.; Zhu, X.N. Environmentally-friendly emulsion-like collector prepared from waste oil: Application in floatation recovery of unburned carbon in coal fly ash. *J. Clean. Prod.* **2022**, *379*, 134561. [CrossRef]
33. Gui, X.; Xing, Y.; Wang, T.; Cao, Y.; Miao, Z.; Xu, M. Intensification mechanism of oxidized coal flotation by using oxygen-containing collector α-furanacrylic acid. *Powder Technol.* **2017**, *305*, 109–116. [CrossRef]
34. Chen, S.; Tang, L.; Tao, X.; Chen, L.; Yang, Z.; Li, L. Effect of oxidation processing on the surface properties and floatability of Meizhiyou long-flame coal. *Fuel* **2017**, *210*, 177–186. [CrossRef]
35. Li, E.; Lu, Y.; Cheng, F.; Wang, X.; Miller, J.D. Effect of oxidation on the wetting of coal surface by water: Experimental and molecular dynamics simulation studies. *Physicochem. Probl. Miner. Process.* **2018**, *54*, 1039–1051. [CrossRef]
36. Shen, L.; Wang, C.; Min, F.; Liu, L.; Xue, C. Effect of pores on the flotation of low-rank coal: An experiment and simulation study. *Fuel* **2020**, *271*, 117557. [CrossRef]
37. Xia, W.; Xie, G.; Ren, C.; Zhang, Z.; Liang, C.; Ge, X. Effect of Natural Weathering Processes on Size and Density Compositions of Bituminous Coal. *Energy Fuels* **2014**, *28*, 4496–4500. [CrossRef]
38. Li, T.; Wu, J.J.; Wang, X.G.; Huang, H. Particle Size Effect and Temperature Effect on the Pore Structure of Low-Rank Coal. *ACS Omega* **2021**, *8*, 5865–5877. [CrossRef]
39. Gupta, A.; Eral, H.B.; Hatton, T.A.; Doyle, P.S. Controlling and predicting droplet size of nanoemulsions: Scaling relations with experimental validation. *Soft Matter* **2016**, *12*, 1452–1458. [CrossRef]
40. Rao, J.; McClements, D.J. Lemon oil solubilization in mixed surfactant solutions: Rationalizing microemulsion & nanoemulsion formation. *Food Hydrocoll.* **2012**, *26*, 268–276. [CrossRef]
41. Kong, M.; Chen, X.G.; Kweon, D.K.; Park, H.J. Investigations on skin permeation of hyaluronic acid based nanoemulsion as transdermal carrier. *Carbohydr. Polym.* **2011**, *86*, 837–843. [CrossRef]
42. Anton, N.; Vandamme, T.F. Nano-emulsions and micro-emulsions: Clarifications of the critical differences. *Pharm. Res.* **2011**, *28*, 978–985. [CrossRef]
43. Amiri, N.; Pourjahed, A.; Karimi, A.S. Preparation and physicochemical evaluation of phytosterolenriched sesame oil microemulsions. *J. Food Process Eng.* **2024**, *47*, e14570. [CrossRef]

44. Tadros, T.; Izquierdo, P.; Esquena, J.; Solans, C. Formation and stability of nano-emulsions. *Adv. Colloid Interface Sci.* **2004**, *108*, 303–318. [CrossRef] [PubMed]
45. Modarres-Gheisari, S.M.M.; Gavagsaz-Ghoachani, R.; Malaki, M.; Safarpour, P.; Zandi, M. Ultrasonic nano-emulsification—A review. *Ultrason. Sonochem.* **2019**, *52*, 88–105. [CrossRef] [PubMed]
46. Perazzo, A.; Preziosi, V.; Guido, S. Phase inversion emulsification: Current understanding and applications. *Adv. Colloid Interface Sci.* **2015**, *222*, 581–599. [CrossRef]
47. Chen, Y.; Mo, S.; Chen, J.; Yu, Y.; Jia, L.; Chen, Y. A novel lauric acid in silicone oil phase change emulsion with enhanced thermal properties and stability prepared by emulsion inversion point method. *Sol. Energy Mater. Sol. Cells* **2024**, *271*, 112855. [CrossRef]
48. Goodarzi, F.; Zendehboudi, S. A Comprehensive Review on Emulsions and Emulsion Stability in Chemical and Energy Industries. *Can. J. Chem. Eng.* **2019**, *97*, 281–309. [CrossRef]
49. Li, W.; Leong, T.S.; Ashokkumar, M.; Martin, G.J. A study of the effectiveness and energy efficiency of ultrasonic emulsification. *Phys. Chem. Chem. Phys. PCCP* **2018**, *20*, 86–96. [CrossRef]
50. Grijalvo, S.; Rodriguez-Abreu, C. Polymer nanoparticles from low-energy nanoemulsions for biomedical applications. *Beilstein J. Nanotechnol.* **2023**, *14*, 339–350. [CrossRef]
51. Pal, R. Shear Viscosity Behavior of Emulsions of Two Immiscible Liquids—ScienceDirect. *J. Colloid Interface Sci.* **2000**, *225*, 359–366. [CrossRef] [PubMed]
52. Kim, K.-M.; Oh, H.M.; Lee, J.H. Controlling the emulsion stability of cosmetics through shear mixing process. *Korea-Aust. Rheol. J.* **2020**, *32*, 243–249. [CrossRef]
53. Escudier, M.P.; Gouldson, I.W.; Pereira, A.S.; Pinho, F.T.; Poole, R.J. On the reproducibility of the rheology of shear-thinning liquids. *J. Non-Newton. Fluid Mech.* **2001**, *97*, 99–124. [CrossRef]
54. Choi, S.B.; Lee, J.S.; Baik, S.J.; Lee, J.S. Numerical simulations for the rheological characteristics of emulsions under several conditions including temperature, shear rate, surfactant concentration and droplet size. *Micro Nano Lett.* **2014**, *9*, 896–900. [CrossRef]
55. Davies, J.T. Drop sizes of emulsions related to turbulent energy dissipation rates. *Chem. Eng. Sci.* **1985**, *40*, 839–842. [CrossRef]
56. Yao, C.; Zhao, S.; Liu, L.; Liu, Z.; Chen, G. Ultrasonic Emulsification: Basic Characteristics, Cavitation, Mechanism, Devices and Application. *Front. Chem. Sci. Eng.* **2022**, *16*, 24. [CrossRef]
57. Leong, T.S.H.; Zhou, M.; Kukan, N.; Ashokkumar, M.; Martin, G.J.O. Preparation of water-in-oil-in-water emulsions by low frequency ultrasound using skim milk and sunflower oil. *Food Hydrocoll.* **2017**, *63*, 685–695. [CrossRef]
58. Ashokkumar, M. The characterization of acoustic cavitation bubbles—An overview. *Ultrason. Sonochem.* **2011**, *18*, 864–872. [CrossRef]
59. Ashokkumar, M. Applications of ultrasound in food and bioprocessing. *Ultrason. Sonochem.* **2015**, *25*, 17–23. [CrossRef]
60. Costa, A.L.R.; Gomes, A.; Cunha, R.L. One-step ultrasound producing O/W emulsions stabilized by chitosan particles. *Food Res. Int.* **2018**, *107*, 717–725. [CrossRef]
61. Feng, J.; Esquena, J.; Rodriguez-Abreu, C.; Solans, C. Key features of nano-emulsion formation by the phase inversion temperature method. *J. Dispers. Sci. Technol.* **2021**, *42*, 1073–1081. [CrossRef]
62. Solans, C.; Izquierdo, P.; Nolla, J.; Azemar, N.; Garcia-Celma, M.J. Nano-emulsions. *Curr. Opin. Colloid Interface Sci.* **2005**, *10*, 102–110. [CrossRef]
63. Anton, N.; Benoit, J.P.; Saulnier, P. Design and production of nanoparticles formulated from nano-emulsion templates—A review. *J. Control. Release* **2008**, *128*, 185–199. [CrossRef] [PubMed]
64. Kupikowska-Stobba, B.; Kasprzak, M. Fabrication of nanoparticles for bone regeneration: New insight into applications of nanoemulsion technology. *J. Mater. Chem. B* **2021**, *9*, 5221–5244. [CrossRef] [PubMed]
65. Feng, J.; Rodríguez-Abreu, C.; Esquena, J.; Solans, C. A Concise Review on Nano-emulsion Formation by the Phase Inversion Composition (PIC) Method. *J. Surfactants Deterg.* **2020**, *23*, 677–685. [CrossRef]
66. Lee, H.S.; Morrison, E.D.; Frethem, C.D.; Zasadzinski, J.A.; Mccormick, A.V. Cryogenic electron microscopy study of nanoemulsion formation from microemulsions. *Langmuir ACS J. Surf. Colloids* **2014**, *30*, 10826–10833. [CrossRef]
67. Gui, X.; Liu, J.; Cao, Y.; Miao, Z.; Li, S.; Xing, Y.; Wang, D. Coal preparation technology: Status and development in China. *Energy Environ.* **2015**, *26*, 997–1013. [CrossRef]
68. Van Netten, K.; Moreno-Atanasio, R.; Galvin, K.P. Fine Particle Beneficiation through Selective Agglomeration with an Emulsion Binder. *Ind. Eng. Chem. Res.* **2014**, *53*, 15747–15754. [CrossRef]
69. Lu, Y.; Wang, X.; Liu, W.; Li, E.; Cheng, F.; Miller, J.D. Dispersion behavior and attachment of high internal phase water-in-oil emulsion droplets during fine coal flotation. *Fuel* **2019**, *253*, 273–282. [CrossRef]
70. Dickinson, J.E.; Jiang, K.; Galvin, K.P. Fast flotation of coal at low pulp density using the Reflux Flotation Cell. *Chem. Eng. Res. Des.* **2015**, *101*, 74–81. [CrossRef]
71. Van Netten, K.; Moreno-Atanasio, R.; Galvin, K.P. Selective agglomeration of fine coal using a water-in-oil emulsion. *Chem. Eng. Res. Des.* **2016**, *110*, 54–61. [CrossRef]
72. Xue, Z.; Dong, L.; Hao, B.; Fan, M.; Chen, W.; Yang, C.; Li, H.; Li, S.; Yang, H. Effect and mechanism of microemulsion on low-rank coal flotation by mixing AEO-3 and diesel oil. *Energy Sources Part A Recovery Util. Environ. Eff.* **2022**, *44*, 4123–4140. [CrossRef]
73. Ruikang, D.; Biyang, T.; Yun, G.T. Synergistic effect of SPS compound surfactant and diesel oil on low-rank coal slime flotation. *Int. J. Coal Prep. Util.* **2023**, *43*, 2091–2105. [CrossRef]

74. Cheng, G.; Zhang, M.N.; Zhang, Y.H.; Lin, B.; Zhan, H.J.; Zhang, H.J. A novel renewable collector from waste fried oil and its application in coal combustion residuals decarbonization. *Fuel* **2022**, *323*, 124388. [CrossRef]
75. Gong, J.; Shen, L.; Liu, Y.; Qiao, E.; Liu, L.; Min, F. Study on the intermediate products of Fischer Tropsch synthesis as a collector for coal flotation: Molecular dynamics and experimental studies. *Fuel* **2024**, *364*, 131125. [CrossRef]
76. Li, E.; Xiao, X.; Wang, X.; Pan, Z.; Qin, Y.; Gao, G.; Du, Z.; Cheng, F. Interfacial interaction of emulsion collector in enhancing low-rank coal flotation. *Colloids Surf. A Physicochem. Eng. Asp.* **2024**, *692*, 133965. [CrossRef]
77. Mcclements, D.J.; Jafari, S.M. Improving emulsion formation, stability and performance using mixed emulsifiers: A review. *Adv. Colloid Interface Sci.* **2018**, *251*, 55–79. [CrossRef]
78. Bensley, C.N.; Swanson, A.R.; Nicol, S.K. The effect of emulsification on the selective agglomeration of fine coal. *Int. J. Miner. Process.* **1977**, *4*, 173–184. [CrossRef]
79. Li, Y.; Xia, W.; Peng, Y.; Xie, G. A novel coal tar-based collector for effective flotation cleaning of low rank coal. *J. Clean. Prod.* **2020**, *273*, 123172. [CrossRef]
80. Li, H.; Meng, Z.; Zhu, G.; Li, S.; Yan, K. A System Device for Removing Impurities from Calcium Carbide Slag: C CN202120471113.5, 9 August 2022.
81. Wang, P.; Liu, W.; Zhuo, Q.; Luo, Q.; Sun, X.; Li, J. Mechanism of interaction between mixed collectors and low-rank coal: An experimental and simulation study. *Int. J. Coal Prep. Util.* **2023**, *43*, 346–360. [CrossRef]
82. Zhang, Q.S.; Niu, C.K.; Bu, X.N.; Bilal, M.; Ni, C.; Peng, Y.L. Enhancement of Flotation Performance of Oxidized Coal by the Mixture of Laurylamine Dipropylene Diamine and Kerosene. *Minerals* **2021**, *11*, 1271. [CrossRef]
83. Ding, L.P. Effect of Collector Interfacial Tension on Coal Flotation of Different Particle Sizes. *Ind. Eng. Chem. Res.* **2010**, *49*, 3769–3775. [CrossRef]

MDPI AG
Grosspeteranlage 5
4052 Basel
Switzerland
Tel.: +41 61 683 77 34

Minerals Editorial Office
E-mail: minerals@mdpi.com
www.mdpi.com/journal/minerals

www.ingramcontent.com/pod-product-compliance
Lightning Source LLC
Chambersburg PA
CBHW062043220926
51850CB00010B/2646

* 9 7 8 3 7 2 5 8 5 6 1 7 6 *